# Unleashing the Power of IT Managed Services: Your Comprehensive Guide

**Dr. Julia Diakakis & Saki Diakakis**

ISBN: 978-1-962557-00-9

# DEDICATION

**This book is dedicated to IT professionals throughout the world.**

# Table of Contents

**Understand Managed Services** .......... 9

**The Ideal Client** .......... 9

When Not to Use Managed Services .......... 10

Managed Services for Increased Productivity .......... 12

Understanding Managed Services in the IT Industry .......... 13

Benefits of Managed Services .......... 14

Increased Productivity .......... 15

Better Decision-Making .......... 16

Improved Security .......... 17

Increased Reliability .......... 19

Improved Scalability .......... 21

Faster Response Time .......... 23

Enhanced Reporting .......... 24

Greater Flexibility .......... 25

**How to Choose the Right Managed Services Provider** .......... 31

Understand Your Needs .......... 31

Ask the Right Questions .......... 32

Consider the Cost .......... 33

Set Expectations .......... 33

Compare Features .......... 35

Obtain References .......... 36

Discuss Details .......... 37

**MSP Benefits Compare to Break-Fix ........................................ 39**

Cost ........................................................................................................39

Security ....................................................................................................39

Expertise ..................................................................................................39

Time Savings ...........................................................................................40

Quality of work .......................................................................................40

Accountability ..........................................................................................41

Warranty ..................................................................................................41

**Core Services Most MSPs and MSSPs Offer ................................ 43**

MS Office (M365) and Variations ..........................................................43

Spam Filtering ..........................................................................................44

Antivirus ...................................................................................................46

Email Archiving .......................................................................................48

Backup .....................................................................................................50

Cloud Solutions ........................................................................................51

Financial Packages ...................................................................................53

Adobe Product Line .................................................................................55

Custom Applications ................................................................................57

**Steps to Implement and Maintain Managed Services .................. 59**

Define Your Needs ...................................................................................59

Discuss Implementation Details ..............................................................60

Research Providers ...................................................................................62

Draft a Contract .......................................................................................62

Contents of an MSP-Client Contract .......................................................63

Implement the Services ........ 64

Monitor Performance ........ 64

SLA (Service Level Agreement) ........ 64

Key Indicators and Metrics ........ 66

**Resolving Issues with Managed IT Services ........ 69**

Understanding System Expiration, Renewals, and Warranties ........ 69

Requesting Regular Status Updates and Establishing a Communication Plan for Notifications ........ 69

Utilizing Remote Access Solutions for Immediate Support When Needed .70

Invest in Cybersecurity Measures to Protect Your Network Against External Threats ........ 71

Keeping Track of Changes in Technology ........ 73

**Advanced Concepts ........ 75**

The Client Portal ........ 75

The LMS (Learning Management System) ........ 77

Document Management System ........ 78

Website ........ 80

WordPress for Websites ........ 82

Appealing MSPs not Providing Support for All Services ........ 83

MSP vs MSSP ........ 84

Quality Assessment ........ 85

MSP & MSSP Together ........ 86

**Equipment Purchases ........ 89**

Questions on Purchasing ........ 89

Leasing Hardware as a Service (HaaS) ........ 90

Software as a Service (SaaS) ..... 91

Financing for Purchasing, and Project Labor ..... 92

**Conflict & Resolution ..... 95**

Manage Conflict & Resolution ..... 95

Conflict Between MSP and Client ..... 96

Changing MSPs ..... 97

Backup the Data ..... 98

Give a Chance to Your Current MSP ..... 99

**Contract and Agreement Guide ..... 101**

Contracts ..... 101

Question of Contract Necessity ..... 101

Contracts & Agreements ..... 102

**Resources to Help Choosing an MSP ..... 105**

Best Practices on Selecting an MSP ..... 105

Benefits of Hiring an MSP Auditor ..... 105

Where to Look to Find a Good MSP ..... 106

Hiring an MSP for a Trial Period ..... 108

**Thoughts on Selecting an MSP ..... 111**

**Biographies ..... 114**

Dr. Julia Diakakis ..... 114

Saki Diakakis ..... 116

# Understand Managed Services

## The Ideal Client

Managed IT services can be ideal for businesses of all sizes. Companies with limited resources and expertise may find that managed services are the most cost-effective solution to managing their IT needs. Similarly, small businesses can benefit from having access to a team of managed service professionals who can provide them with more comprehensive IT support than they would be able to afford on their own. Additionally, larger enterprises can take advantage of the scalability and flexibility that managed IT services offer as well as benefit from the access to newer technology and experience offered by third-party providers. Managed services provide businesses with better security, reliability, and performance while helping cut costs and free resources for other important tasks.

If your business relies on IT services, such as emails, websites, shared folders, workstations, and printers to function effectively, along with corporate backups for workstations and servers - consider partnering with a managed service provider. With the right partner, the organization can benefit from reliable performance at optimal cost efficiency to include on-boarding experienced personnel to assist in setup & configurations plus providing security solutions tailored specifically to your company's needs. If your organization has reached the point wherein comprehensive IT services and support is needed, but footing the bill for a committed team is not feasible, then this is the perfect solution for you! This does not mean that companies with deep enough pockets to employ full-scale IT teams are not eligible. With managed service providers on-boarding enables access to quality of service and features at an optimal cost while guaranteeing availability across all systems.

## When Not to Use Managed Services

Businesses that prefer to control their own IT infrastructure and networks should not use managed services. Additionally, businesses with a well-defined IT strategy may find that they can manage their network more efficiently in-house. For these businesses, the cost of outsourcing may outweigh the benefits provided by managed services. It is important to weigh the advantages and disadvantages of managed services against those of in-house management to determine which solution is beneficial for a particular business. Finally, companies whose IT operations are heavily regulated should also consider carefully before outsourcing IT management and maintenance as this may complicate compliance with certain regulations.

Managed services can be an effective solution for businesses of all sizes to simplify and improve the management of their IT infrastructure. However, it is important to assess your business's needs and capabilities to determine whether managed services will provide the most cost-effective and efficient solution. Consulting a professionally managed service provider is a terrific way to get advice on how best to approach IT management for your organization. Whether you decide to use managed services or manage your network in-house, you must have a dependable team of professionals monitoring your systems to ensure optimal performance and security.

If your company is not in a strong economic position, then managed services should be avoided due to the risk of jeopardizing operations and putting additional strain on staff. Only invest in support for features that will make your business more profitable or reduce workloads, without going beyond what is realistically affordable. Utilizing this strategy guarantees that your services will remain working seamlessly - a feat impossible to accomplish otherwise. Furthermore, managed services provide a comprehensive suite of tools and resources that can be used to keep your business running at its best.

If your business relies on specific manufacturers for hardware, software, or applications such as custom services, then a managed service provider may not be the best fit since most of them focus solely on their supported solutions. Of course, some providers will make exceptions to this rule--but that usually comes at an added cost. Be sure to consider any hidden fees if you require support beyond what they offer in their regular operating mode.

Summary list:

- Your company is not financially strong.
- You do not like IT engineers and specialists changing often.
- You like having control over the tech stock used (software for backups, email, etc.)
- You would like to have full control over the response times and outcomes.
- You do not feel comfortable with external resources managing company assets.
- You value privacy and your customer's privacy beyond the regulatory methodologies.

# Managed Services for Increased Productivity

Managed services can offer business owners a wide range of benefits, including access to sophisticated technologies and experienced personnel. Businesses with limited resources and small IT budgets may find that managed services are the most cost-effective solution to managing their IT needs.

Managed services cover a range of different IT services, each presenting its challenges and complexities. For example, cloud computing requires an understanding of server-side security, while network monitoring involves keeping track of increasingly complex systems with fewer resources. Security measures need to be regularly updated to protect networks from cyber threats, while data centers must be managed efficiently to ensure reliability and performance.

The need for managed services is also driven by advances in technology. As new tools emerge, businesses must continually update their tools and strategies to stay ahead of the competition and remain secure. Managed services help ensure that businesses have access to the latest technology, as well as the personnel who specialize in managing it effectively. The unique areas of this industry include the following:

Cloud Computing: providing secure cloud infrastructure, storage solutions, automation, and analytics.

Network Monitoring & Security: monitoring networks for security vulnerabilities, unauthorized activity, and performance issues; setting up firewalls and other measures to reduce attack surface area; installing malicious software detection tools; scanning for malware; etc.

Data Center Management & Automation: managing servers and networks within data centers; automating processes such as provisioning or deployment; setting up governance frameworks for maintenance tasks; etc.

Artificial Intelligence (AI) & Machine Learning: using AI techniques to analyze large datasets for strategic decision making; building chatbots or virtual assistants; leveraging natural language processing (NLP); implementing machine learning algorithms for automated functions like fraud detection or customer service management; etc.

## Understanding Managed Services in the IT Industry

Managed services in the IT industry refer to outsourcing the maintenance, management, and monitoring of an organization's IT infrastructure and networks to a third-party provider. By partnering with a managed service provider (MSP), businesses can access the latest technologies and have a team of professionals monitoring and managing their networks and systems continually. This ensures that any potential risks to the business's IT operations are minimized, allowing the business to focus on generating revenue instead of worrying about IT problems. Managed services also provide organizations with improved scalability and flexibility in terms of customizing their IT solutions, as well as improved cost predictability for budgeting purposes.

# Benefits of Managed Services

To restate the benefits of managed services, there is the provision of cost-effective IT solutions for all business types and sizes. Outsourcing IT operations to a managed service provider, allows access to technologies and experienced professional teams. These teams are capable of monitoring and managing networks and systems. This ensures minimization of security risks, allowing revenue generation to be the primary focus of the organization. Lastly, predictability of costs associated with outsourcing and increased operational visibility are added benefits.

## Cost Savings

Computer-managed services can minimize costs by reducing physical labor, energy, and other resources required to manage large workloads.

Computer-managed services offer significant cost savings compared to manual processes. Automation can reduce physical labor costs, energy costs, and other resources used in the management of workloads. It also helps companies save time as mundane tasks such as data entry are automated, allowing personnel to focus on more important initiatives. Furthermore, automated security systems can help protect sensitive information while providing access control measures that ensure only the right people have access to certain parts of the network. Computer-managed services provide cost savings through the efficient use of resources and reduced overhead costs associated with manual operations.

By controlling your equipment costs, you can enjoy the maintenance and tax benefits of Hardware or Software as a Service (HaaS, SaaS). Not only does this save money on staffing IT personnel, but you will also be able to procure better deals for some products than if purchased individually. Additionally, with 24/7 support when necessary and budget planning without needing additional financial resources or data analyst tools, it is an efficient way to assess company workflows in terms of their relationship with IT.

## Increased Productivity

Computer-managed services offer a variety of benefits to organizations, including increased productivity and cost savings. By eliminating manual labor costs, automating mundane tasks, and providing access control measures, businesses can reduce overhead costs associated with IT operations and focus on more important initiatives. Automation also allows for faster data analysis, allowing businesses to make informed decisions quickly. Additionally, automated systems are often able to detect errors quicker than human operators, leading to improved customer service and satisfaction. Furthermore, computer-managed services provide an efficient way for businesses to manage their resources; predict costs; increase visibility into their overall IT operations; and improve scalability in terms of customizing IT solutions to suit specific needs.

In addition, managed services allow organizations of all sizes to leverage natural language processing (NLP) and machine learning algorithms to automate functions such as fraud detection or customer service management. This can give companies the ability to easily implement strategic decision-making processes while building chatbots or virtual assistants that improve customer experience by responding quickly and accurately to queries. These factors contribute towards increased efficiency when it comes to managing workloads, resulting in significant cost savings as well as improved customer service levels over time.

# Better Decision-Making

Computer-managed services allow organizations to make better decisions faster. Automated systems can rapidly process substantial amounts of data and analyze it in an accurate and unbiased manner, providing the insights needed for informed decision-making. Automation eliminates human biases that may otherwise influence the decision-making process, resulting in more logical and rational decisions. Additionally, automated systems can store historical records of previously made decisions, allowing business leaders to draw on past experiences to make better-informed choices. Automated decision-making processes have significantly improved organizational efficiency by streamlining the decision-making process with greater accuracy and consistency.

Leveraging computer-managed services to manage IT networks and systems help organizations reduce costs, increase productivity, and improve decision-making. By automating mundane tasks such as data entry and analysis, companies can save money on labor costs while gaining access to insights that would otherwise be unavailable. The use of computer-managed services has allowed businesses to better utilize their resources and make more informed decisions to remain competitive in today's market.

By utilizing Key Performance Indicators (KPIs), metrics, and key indicators organizations can take advantage of these cost savings and efficiency improvements offered by computer-managed services while also gaining insight into how their operations are performing. By focusing on these KPIs, metrics, and key indicators organizations can set realistic goals, measure progress towards those goals, and make informed decisions over the long term that will help improve efficiency, reduce costs, and increase productivity. Setting up KPIs, metrics and key indicators allows organizations to monitor performance in real time which helps ensure that processes are continually improved, and resources are allocated more effectively. Through this improved decision-making process, companies can stay ahead of their competitors and remain competitive.

Computer-managed services provide cost savings through efficient utilization of resources as well as increased productivity by automating tedious tasks such as data entry and analysis. Additionally, KPIs, metrics, and key indicators can help organizations make informed decisions over the long term by allowing them to monitor performance in real time and set realistic goals that will improve efficiency, reduce costs, and increase productivity. By leveraging these tools companies can remain competitive in today's market.

## Improved Security

Automated security systems can be used to protect information from malicious actors while simultaneously providing access control measures that ensure only the right people have access to sensitive parts of your network.

Computer-managed services offer improved security compared to manual systems. Automated systems allow organizations to quickly detect unauthorized access and respond appropriately, preventing malicious actors from exploiting potential vulnerabilities. Additionally, automated systems can provide more secure methods of authentication and authorization than traditional paper-based methods. Advanced encryption protocols ensure that sensitive data remains safe and secure while access control allows organizations to fine tune access levels for different personnel or applications. The implementation of automated security measures provides organizations with an increased level of protection from any external threats.

Computer-managed services also offer improved scalability, enabling organizations to quickly and easily increase their network capacity to meet changing demands. This saves time and money, as it eliminates the need for manual upgrades or additional hardware purchases. Additionally, automated security measures can be programmed into a system which provides an extra layer of protection against potential threats. Automating security measures such as two-factor authentication, advanced firewalls, and secure access protocols not only makes networks more secure but also improves the user experience by making login processes simpler and more efficient.

Furthermore, computer-managed services provide organizations with a greater level of flexibility in terms of customization. By utilizing KPIs, developers can customize existing systems to meet specific requirements and integrate them with other applications or platforms if necessary. This allows for faster development times with fewer resources while also improving efficiency and reducing costs associated with manual labor. Additionally, certain software packages offer automatic updates which ensure that your systems are always up-to-date and running on the latest version of the software.

Overall, leveraging computer-managed services benefit organizations significantly in terms of cost savings, improved efficiency, increased productivity, enhanced security, greater scalability, and higher levels of customization compared to manual processes. While having access to these tools is beneficial, it is important to note that they have been designed with security at the forefront of their functionality. Automated security measures such as two-factor authentication protocols make networks more secure while access control ensures that only authorized personnel have access to sensitive areas. Furthermore, advanced encryption technologies help keep data safe from malicious actors while providing access control measures for administrators. With all these benefits combined into one package organizations can be sure that their data remains secure while at the same time increasing efficiency and reducing costs associated with traditional IT management strategies

## Increased Reliability

Computers can manage substantial amounts of data with precision and accuracy, making them highly reliable for managing complex tasks.

Computer-managed services offer increased reliability over manual systems. Automated processes allow for greater accuracy in operations, eliminating the potential for human error. Additionally, automated systems can detect and respond to potential system glitches and bugs quickly and efficiently, avoiding costly downtime in the process. Automation also allows organizations to proactively identify incoming threats before they can cause any damage, further increasing overall reliability. Automated systems provide organizations with greater control and stability in their operations, ensuring that their data always remains secure and dependable.

Computer-managed services also offer improved scalability, enabling organizations to quickly and easily adjust their network capacity to meet changing demands. This can save time and money, as it eliminates the need for manual upgrades or additional hardware purchases. In addition to this, automation provides organizations with greater control over their operations by allowing them to customize existing systems based on specific requirements and integrate them with other platforms if needed.

Advanced encryption technologies provide another key benefit of computer-managed services, ensuring data remains secure from malicious actors. Automated encryption processes provide a secure and reliable way of protecting confidential information compared to manual methods. Additionally, certain software packages come with automatic updates which ensure that systems are always running on the latest version of the software, reducing the risk of outdated versions being used and making them less vulnerable to external threats.

Automation helps reduce costs associated with manual labor while improving efficiency by eliminating potential human errors. Using computers to automate processes such as data entry or customer service inquiries, organizations can quickly process substantial amounts of data with greater accuracy than is achievable manually. This allows for faster development times with fewer resources while simultaneously improving productivity levels and minimizing expenditure on human labor.

Overall, leveraging computer-managed services has many advantages ranging from cost savings, improved efficiency, and increased productivity to enhanced security and higher levels of customization compared to manual processes. Automation not only provides organizations with an increased level of protection against external threats but also allows for scalability in terms of adjusting their network capacity according to changing demands quickly and easily without having to purchase additional hardware or manually upgrade systems. Automated security measures such as two-factor authentication protocols make networks more secure while access control ensures that only authorized personnel have access to sensitive areas thus providing better integrity for users' data. Advanced encryption technologies help keep data safe from malicious actions while providing access control measures for administrators resulting in increased reliability no matter what changes occur in an organization's environment.

## Improved Scalability

Automation allows for the easy scaling of infrastructure and services when more resources are needed.

Computer-managed services offer improved scalability thanks to automation and robust infrastructure. Automation simplifies the deployment of new services, allowing organizations to quickly scale up as needed with minimal downtime. Furthermore, automated systems can automatically reconfigure resources as needed in response to changing demand, ensuring that they remain optimally utilized always. Automation and improved infrastructure enable organizations to rapidly expand their operations without compromising security or reliability.

Computer-managed services offer other benefits, further enhancing overall utility. For example, automation allows for increased customization and flexibility when managing complex tasks. Administrators can easily set up custom scripts to automate the most time-consuming and repetitive processes, allowing them to focus on more important tasks while still having the data they need available whenever they require it. This increased level of control also enables users to tailor their systems according to their specific needs, such as setting thresholds or triggers for automated actions.

In addition, many computer-managed services come with built-in monitoring features enabling administrators to monitor system performance in real time with real-time metrics reporting. This provides better visibility into performance issues and helps administrators quickly identify potential problems or areas of improvement before they become serious issues. Furthermore, automation allows for faster detection of security breaches and enables administrators to take immediate action in response. Automated software patches also allow organizations to ensure that their systems are always running the latest version of the software without any manual intervention required.

Moreover, computer-managed services offer a greater level of reliability and consistency as well as improved scalability due to automation and robust infrastructure. Automation simplifies the deployment of new services, allowing organizations to quickly scale up as needed with minimal downtime. Furthermore, automated systems automatically reconfigure resources as needed in response to changing demand which ensures that they remain optimally utilized. Robust network infrastructure allows for seamless scaling with no need for manual intervention or complex programming thus enabling organizations to rapidly expand their operations without compromising security or reliability.

In conclusion, computer-managed services have numerous advantages over traditional manual methods including cost savings, improved efficiency, increased productivity, and enhanced security features. Automation can be used to customize systems according to an organization's specific needs while advanced encryption technologies provide a more secure way of protecting confidential information compared to manual methods. Moreover, automated monitoring features enable administrators to track system performance while automated software patches help ensure that systems are always running on the latest version of the software reducing the risk of outdated versions being used and making them less vulnerable to external threats. Finally, computer-managed services offer improved scalability thanks to automation and robust infrastructure resulting in increased reliability in always managing complex tasks regardless of changes in an organization's environment.

## Faster Response Time

Automated systems can respond to requests or changes faster than manual operations, allowing businesses to act quicker and be more agile.

Computer-managed services offer faster response time compared to manual systems. Automated processes enable organizations to reduce the time it takes to detect and respond to issues, thus allowing them to resolve incidents quickly and efficiently. Additionally, automated systems can identify potential problems before they can cause any damage or disruption, providing organizations with an initiative-taking approach to avoiding costly downtime.

Automation also reduces complexity in deployments, enabling organizations to provision resources faster and get services up and running more quickly. Automated systems provide organizations with greater reliability and control over their operations, ensuring that their data remains secure while offering faster response times than manual systems. As an illustration, if a server is offline, it could take several hours to comprehend that this may hinder printing or the website being down and the processing of orders ceasing. Automated and managed systems with fast response times not only enable you to know immediately but more importantly your IT team will be aware as soon as possible.

With the help of a Managed Service Provider, you can quickly detect and address any issues that may arise in your business. Automation allows for services like rebooting servers automatically, sending notifications, applying patches, or detecting fraud right when it is needed most. An excellent provider will take proactive steps to prevent disruptions from occurring by addressing the problem before anyone knows about it.

## Enhanced Reporting

With automated tracking and reporting, computer-managed services can generate detailed reports that help identify trends and problems quickly.

Automated tracking and reporting of computer-managed services provide organizations with valuable insights into their operations. Automated systems continuously monitor service performance and resource utilization, providing detailed information on system health and up-time. Additionally, automated tracking and reporting systems can generate detailed reports that identify anomalies or trends in usage. These reports can be used to make informed decisions about resource allocation or to quickly detect and mitigate issues before they become critical. Automated tracking and reporting systems also offer granular visibility into a system's operations, allowing organizations to make changes quickly while ensuring that they remain compliant with industry regulations. Automated tracking and reporting provide organizations with the information they need to keep their services running reliably and efficiently.

IT reporting can have a significant impact on business flows providing organizations with detailed insights into their operations. Automated tracking and reporting systems can provide granular visibility into service performance and resource utilization, allowing organizations to quickly identify potential problems or anomalies. This not only helps organizations address issues before they become critical but also enables them to make informed decisions about resource allocation to ensure maximum efficiency. Additionally, automated tracking and reporting systems can generate reports that identify trends or patterns in usage, allowing businesses to recognize areas of opportunity or improvement more quickly. In addition to improving operational flow, automated tracking and reporting systems help organizations maintain compliance with industry regulations by providing detailed information on system health and up-time. With the help of IT reporting, businesses can improve their operational flows by ensuring that their data remains secure while making informed decisions about resource allocation and recognizing potential.

## Greater Flexibility

Automation makes it possible to add new features and functions quickly, making applications and services more flexible.

Computer-managed services offer organizations increased flexibility and scalability in their operations. Automated systems can quickly detect changing conditions and modify service settings, accordingly, enabling organizations to better meet customer demands. Additionally, computer-managed services can be deployed across multiple environments, allowing organizations to scale up or down quickly and efficiently as needed. Automation also allows organizations to implement automated deployment processes, eliminating the need for manual intervention and providing faster time-to-market for new products or services. Computer-managed services provide organizations with the agility they need to successfully navigate ever-changing market conditions and stay competitive in their field.

# Solutions Offered through Managed Services

Managed services are increasingly becoming a preferred option for businesses of all sizes as they provide benefits such as improved performance, greater scalability and agility, cost savings, and higher reliability.

Managed services are an invaluable tool for businesses looking to stay up to date with the latest technology and developments while ensuring their IT infrastructure is secure, reliable, and effective. With both automated systems and managed services working together, businesses can reduce costs associated with IT projects, ensure data security and up-time, increase employee productivity, manage customer interactions more effectively, and ultimately stay competitive in the marketplace. Automated systems provide the necessary control and flexibility while managed services.

As previously stated, automated systems provide organizations with greater control and flexibility. Automated systems detect changing conditions and modify service settings, allowing for faster response times and improved performance. Additionally, automated tracking and reporting systems generate detailed reports that identify trends or anomalies. Automated systems also provide organizations with greater scalability and flexibility, allowing them to quickly add new features and functions. Furthering to the advantages of the offerings of automated systems is the power and control needed to stay competitive in the digital world.

Managed services provide businesses with an extensive range of solutions to help increase their efficiency, competitiveness, and security. Solutions vary depending on the company providing them, but include:

**Network Management:**
Setting up networks such as LANs and WANs, configuring security measures, and managing performance.

**Data center management:**
Ensuring reliable up-time, monitoring network health and performance, patching software updates, optimizing resource usage, and capacity planning.

**Infrastructure virtualization:**
Creating virtual environments for applications or servers to improve scalability, flexibility, cost savings, and security.

**Cloud computing & storage:**
Providing secure cloud infrastructure for data storage and applications development; enabling automation features like auto-scaling; etc.

**Security & compliance:**
Installing firewalls or other security measures; carrying out penetration testing; meeting compliance requirements (e.g., GDPR, NIST 800-171, MCCM); etc.

**Business continuity & disaster recovery planning (BCDR):**
Ensuring readiness in case of system failure or unexpected downtime; implementing incident response plans; rehearsing recovery scenarios; etc.

**Telephony:**
With modern phone systems such as VOIP, various tools can be used to maximize your telephone service.

Automated call queues and alerts assist in workflow management for customer inquiries. Special messages for events or holidays allow you to customize interactions with clients and voice messaging assists in managing incoming communications quickly and efficiently.

**Security surveillance technologies:**
Installing and managing video surveillance systems for security purposes.

**Monitoring & reporting:**
Collecting data on system performance, usage, up-time, and other metrics; analyzing the data to identify issues or trends; providing regular reports that can help inform business decisions. Managed services offer organizations a comprehensive and reliable solution for managing their IT needs.

**Productivity applications:**
Managing and supporting a wide range of applications, such as productivity suites (e.g., Office 365), customer relationship management systems (CRM), project management software, etc., to help teams work more efficiently.

**Finance applications:**
Setting up and managing software for financial operations. This includes accounting systems, analytic platforms, payroll and HR software, invoicing tools, etc. provides the expertise and support required to keep a business running smoothly and efficiently. With the help of managed services, businesses can focus on their core competencies while receiving reliable IT support from a trusted provider.

**Fire alarms and facility security:**
While not all traditional Manage Services providers deliver these types of services, some do. Installing and

managing fire alarms and other security systems is a key part of ensuring the safety of staff and customers, as well as protecting physical assets and data. Fire alarm monitoring services help ensure that buildings are protected from potential danger and provide real-time alerts in case of an emergency. Monitoring access control systems helps secure entrances to a building and limit access to specific areas.

# How to Choose the Right Managed Services Provider

## Understand Your Needs

Before you look for a managed services provider (MSP), it is important to understand what your business needs are and how they can be best met. What type of infrastructure and applications do you require? What level of support is needed?

Depending on your industry, certain types of support will be integral to the successful functioning of your business. Take manufacturing as an example; cloud operations, virtual servers, and other remote solutions are essential in this field. Meanwhile, the Health Insurance Portability and Accountability Act (HIPAA) compliance calls for heightened security measures responding to record-keeping – which is why medical organizations often opt out from having their data stored online and instead use on-premises servers and workstations. Although cloud solutions can provide an equal level of security, the decision to choose one over the other is mostly based on cost analysis and individual preference.

When considering the right setup for your business, it is critical to understand that both service and profitability are essential elements of any MSP provider. The number of resources required, type of support needed (24/7 or 8-5 only), and other factors will all influence how much you can expect to pay per device, server, and workstation in terms of your overall budget.

Continuous assessment is key when weighing these components to ensure continuity across all areas of your business.

## Ask the Right Questions

Once you have a clear idea of what your needs are, it is important to ask the right questions when vetting potential providers. Are they experienced in collaborating with companies of your size or industry? Do they have any certifications or specialties? What is their response time for customer inquiries and support requests?

Before engaging with a provider or making a final decision, it is imperative to comprehend and authenticate the significance of the services needed. Is your e-mail system crucial for doing business? Could you go without an e-mail for hours or days if necessary? How would losing access to company files affect operations over a day or two? Does your organization rely heavily on print services such as work orders, labels, and envelopes being printed daily? What might be the repercussions of not having these printing capabilities even just briefly? Be sure you fully understand the needs of your business while vetting providers.

Reach out to your MSP with the entire understanding of what you need for your business to be successful. Make sure they comprehend all the fundamentals and needs, as well as any ideal conditions that would help make businrss run smoother. Ensure that they understand every bit of information regarding your environment so they can deliver a service tailored specifically to you.

## Consider the Cost

Cost can be a major factor when selecting a Managed Services Provider. It is important to compare options from different providers and make sure that you are getting the best value for your money. In addition to looking at upfront costs, also consider any associated fees for additional services or maintenance.

The cost of business continuity is not one to be taken lightly, as it can put a strain on corporate resources. However, the quality of service received more than makes up for price - there is a significant cost difference between an hour response and twenty-hour response time; real-time fail-over capabilities versus backup & restore methods; or slow backups with fast restores. Investing in these services now will save your company significantly later! If you have a vision of where you want your business to be and the cost differential between your options, it could potentially determine the success of future growth.

## Set Expectations

When choosing a provider, it is important to set expectations for the relationship. Establishing clear communication regarding the scope of services and expected outcomes will help ensure smooth operations down the road. Having a provider that offers flexibility also allows for business need changes over time.

Using Service Level Agreements (SLAs) when selecting an MSP is essential in getting the best value for your investment. With an SLA, you can set clear expectations regarding response times and priority levels, gain access to tailored packages and scalability options, and ensure that your business needs are met with quality service. It is important to ask the right questions and compare costs before choosing a provider, as well as evaluate their experience so that you can make the most informed decision possible. With the right managed services provider on your side, you will be able to focus more on achieving success rather than dealing with costly issues down the road.

Quality of service is a crucial factor when selecting an MSP. Without setting expectations and having a clear understanding of the scope of services, you risk delivering sub-par or insufficient customer service. Make sure to compare providers and look for one with experience in your industry, as well as flexible solutions that can accommodate changing business needs. With the right provider on your side, you can rest assured that quality service will be delivered, and your business will be safe.

Setting expectations for the quality of service is critical for ensuring success both now and in the future. Taking the time to consider factors such as cost, scalability options, response times, priority levels, and tailored packages can improve customer satisfaction while helping your business to run more smoothly. Quality service is an investment that can pay off eventually and help you achieve maximum success.

The right MSP, coupled with clear expectations for quality of service, will help ensure a smooth and successful business relationship now and into the future.

## Research Options

Take the time to research different providers and compare shops. Read customer reviews and speak with other businesses that have used a managed services provider for similar projects. This will help you find the best fit for your business needs.

You should also consider cost and scalability options, as well as experience in your industry. Additionally, evaluate the different sets of services offered to ensure that you are getting the most value for your money. Lastly, make sure to compare customer service response times, priority levels, and tailored packages when selecting an MSP - this is essential for ensuring quality customer service.

By taking the time to research potential providers and evaluate their offerings carefully, you can rest assured that you will be making an informed decision that best suits your business needs. Investing in quality managed services now can save time and money down the road while helping to keep your customers satisfied with timely and reliable service.

Taking these steps to find the right MSP will set your business up for success now and in the future. With a quality provider on your side, you can focus more of your time and resources on achieving success.

## Compare Features

Once you have narrowed down your list of potential providers, compare their features side by side. Look at things such as security measures, backup and recovery capabilities, scalability options, customer service response times, and any additional services that may be included in the package. As you plan, research services such as telephony, surveillance, and facility security. Assess which provider has a strong financial footing to cover potential growth. A seasoned company with a proven record of successful methodologies is key in making an informed decision when choosing your service provider.

It is essential to understand the financial stability of a service provider before you hire them. Determine if they operate from an office space or home and try to request a personal tour around their facility so you can assess how your data will be overseen. Seeing the actual infrastructure in person will help ensure that all questions are answered about their services and provide peace of mind during the selection process.

Does the provider have a full-time data analyst? This is an incredibly essential factor to consider. Most professional automation systems offer rudimentary analysis, but if you want accurate insights into your business for future decisions, then analytics that go beyond the basics – such as trends, patterns in behavior, and important metrics – are critical. Propelling your business to new heights through informed decisions that entail software, hardware, and personnel investments is far more valuable than just relying on a set of data. This will help you determine which provider is best suited for your business needs.

## Obtain References

Once you have identified a few potential providers, ask them for references from clients, previous and existing. This will give you an idea of the provider's level of expertise and customer service. Speaking to past clients can provide valuable insight into the quality of service they received.

By taking the time to research your options, thoroughly vet potential providers, and set expectations before signing a contract, you can make sure that you are choosing the right managed services provider for your business. Investing in a quality service provider, assures that your company's IT systems are well-protected and optimized for peak performance. This will help ensure that your business is well positioned for success overall.

Obtaining references from potential MSPs is an essential step in making an informed decision. Doing so helps to guarantee that you are investing in a quality provider with proven experience and a record of satisfied customers. It allows you to research their customer service levels, security measures, backup and recovery capabilities, scalability options, and other included services in more detail before signing any contracts. It also gives you the ability to get a better understanding of how they operate by speaking with past clients and getting a personal tour around their facility.

By procuring references from prospective managed services providers, you can ensure that the provider has a solid financial footing and the capability to meet your current needs as well as any future growth. Not only does this provide peace of mind but it also gives you valuable insight into their methodologies and expertise. Furthermore, if the provider has access to a full-time data analyst, it can help you make informed decisions when selecting software, hardware, personnel investments, and other critical components for your business.

Taking the time to get references from prospective managed services providers before signing any contracts, provides confidence that your IT systems will be well-protected and optimized for peak performance now and in the future. This will give your company an added advantage ensuring that all necessary steps have been taken toward achieving success.

## Discuss Details

Before signing a contract, discuss the details of what services will be provided and the associated costs. Make sure that you understand all service level agreements and any additional fees for extra services or maintenance. This is also an opportunity to ensure that both parties agree about expectations for communication, service coverage, problem resolution times, and more.

Choose wisely. Finally, trust your gut. Make sure that

you get a good sense of the managed services provider's culture, capabilities, and customer service before signing on the dotted line. Taking the time to do your research will help ensure that you are making an informed decision that is best for your business eventually. Doing this will pay dividends in the future as you move forward with a managed services provider that meets your business needs.

By following these tips, you can make sure that you choose the right managed services provider for your business. Investing in quality service will help ensure that your company's IT systems are secure and optimized for peak performance, setting your business up for long-term success.

# MSP Benefits Compare to Break-Fix

## Cost

The cost savings of utilizing a Managed Service Provider (MSP), compared to break-fix methods, are significant and can help any organization save valuable resources. Having an MSP proactively monitor the health of a network or system allows potential issues to be identified and resolved quickly, preventing costly downtime. As a result, it is easier for businesses to budget for IT services allowing them to take advantage of the cost savings associated with scalability.

## Security

One of the greatest benefits of MSPs is the ability to provide enhanced security for businesses. With the help of advanced tools, an MSP monitors a network and system for threats in real time and acts quickly if malicious activity is detected. This reduces the risk of data breaches or other security incidents that have devastating consequences for companies, such as financial losses, reputation damage, and compliance issues.

## Expertise

When selecting an MSP, access to their combined expertise and experience is gained. Many businesses are unable to afford specialized IT staff with the technical knowledge needed to responsibly manage a network or system. An MSP, however, can provide the expertise of knowledgeable and experienced professionals who are skilled in providing proactive monitoring and troubleshooting services. This ensures that technical issues that arise are dealt with expeditiously and effectively.

## Time Savings

Choosing an MSP reduces IT-related costs in terms of time savings as well. Because of the advanced tools available to them, an MSP can often detect potential issues before they become a critical problem. This enables them to take immediate action in resolving problems quickly and more efficiently than if businesses were relying on break-fix methods for IT needs. This saves valuable resources wasted on unnecessary repairs and allows organizations to focus on other important tasks.

## Quality of work

When utilizing an MSP, businesses can benefit from the quality of work provided in comparison to break-fix. An MSP will have skilled and experienced professionals who can provide industry-leading service when it comes to IT maintenance, support, and troubleshooting. Access to advanced tools helps to ensure that any technical problems are addressed promptly so that businesses do not suffer from costly downtime or worse. As a result of their expertise, an MSP is not only able to resolve existing problems but also proactively monitor networks and systems for any threats, allowing them to reduce the chances of security incidents occurring in the first place. Additionally, by having an MSP manage IT services, businesses can enjoy greater scalability in terms of cost savings while still getting high-quality work performed on time with minimal disruption. All these advantages make an MSP a highly valuable option for any business looking for reliable IT solutions.

## Accountability

Accountability is a crucial factor to consider when choosing an MSP over a break-fix model. Working with an MSP can provide businesses with the assurance that their IT services are being managed in a timely and professional manner while maintaining the highest level of security. Advanced monitoring tools detect potential threats quickly and proactively address any issues before they become costly problems. This provides businesses with greater confidence in their IT systems, knowing that their network and system security is being actively always monitored. Additionally, because most MSPs offer contracts that include service-level agreements (SLA), businesses can hold them accountable for any delays or issues that might arise during their engagement. This helps to ensure that any problems are resolved quickly and efficiently without disruption to normal operations or further costs incurred by the client. Furthermore, having accountability from an MSP ensures that businesses have access to expert advice whenever needed and receive prompt attention when dealing with technical issues or making changes to their networks or systems. All these advantages make an MSP a highly valuable option for any business looking for reliable IT solutions with greater accountability than break-fix models offer.

## Warranty

When choosing an MSP versus a break-fix model for managing IT services, businesses benefit from the added advantage of having a warranty. Working with an MSP provides organizations with peace of mind knowing that their systems and networks are covered in case of any unforeseen issues or technical problems. This not only helps to reduce the risk of costly downtime but also assures that any potential issues will be promptly addressed.

Having a warranty on IT services helps to ensure that businesses have the necessary resources to resolve problems quickly and efficiently without disruption to their operations or further costs incurred by the client. When working with an MSP, businesses benefit from warranties on hardware, software, and service-level agreements. These warranties often include coverage for all aspects of IT management, including installation, maintenance, and repair, as well as support for emerging technologies such as cloud computing solutions.

In addition to protecting against unexpected costs due to system downtime or other technical issues, warranties provided by MSPs also offer protection from manufacturer defects and provide access to replacement parts should any malfunctions occur. This helps organizations avoid costly repairs or replacements since most manufacturers offer limited warranties on their products. Furthermore, warranties offered by MSPs often cover labor costs associated with repairs and replacements so organizations do not have to worry about these additional expenses which could otherwise become too expensive over time.

Overall, having a warranty when working with an MSP provides organizations with greater peace of mind knowing that IT services are being managed in a timely and professional manner while maintaining the highest level of security thanks to advanced monitoring tools. These tools detect any potential threats quickly and proactively address them before they become costly problems. With the added benefit of cost savings due to labor coverage, businesses enjoy greater scalability while receiving high-quality work performed on time with minimal disruption.

# Core Services Most MSPs and MSSPs Offer

## MS Office (M365) and Variations

Microsoft Office applications are a key product offering that help business customers improve their productivity and efficiency. Microsoft Office is one of the most popular productivity suites and is used by millions of professionals around the world. It contains a wide range of applications, from email clients such as Outlook to document and spreadsheet programs such as Word and Excel. The suite also includes presentation software (PowerPoint), database programs (Access), and other useful tools (OneNote and Skype).

MSPs can offer customers a variety of solutions when it comes to supporting their Microsoft Office needs. For those needing basic setup and support, MSPs can provide installation assistance, troubleshooting guidance, and regular updates to keep customers' systems up to date with the latest version of the software. For larger organizations needing more advanced solutions, such as custom configuration or integration with other services, MSPs can tailor specialized solutions.

The importance of using Microsoft Office in today's businesses cannot be underestimated. Using Office applications allows businesses to stay connected with customers, colleagues, partners, and vendors across the globe easily. The suite provides valuable analysis capabilities with its powerful spreadsheet program,

Excel, which allows users to quickly create data sets for decision-making purposes or generate reports for project management. Additionally, its integrated collaboration features allow teams to work together on documents from anywhere in the world at any time. This not only improves workflows but also keeps team members productive even when working remotely.

In today's digital age where data is king, having access to powerful yet user-friendly analytics solutions provided through Microsoft Office is essential for businesses who want to stay competitive. Through the suite's various components - such as Word's mail merge feature - businesses can easily create personalized emails or other documents without having to manually input data into each one individually; this saves time while helping companies become more efficient with their daily tasks.

Overall, Microsoft Office provides businesses with a comprehensive suite of tools that helps them stay connected and organized while improving overall productivity levels within the organization - making it an ideal choice for any company looking for reliable office software solutions for their organizational needs. With an experienced MSP providing top-notch support services, companies are sure to get optimal results out of their investments in Office applications while avoiding IT issues down the line as well.

## Spam Filtering

Spam filtering as part of an MSP's product offering is essential for businesses today as email remains one of the most common forms of communication. By integrating spam filtering applications, businesses can be assured that their incoming communications are safe and secure from malicious threats such as malware, phishing attempts, and other dangerous intrusions. As an MSP providing these services, it is important to sell not only the technology and applications but also the peace of mind that comes with having a secure email system.

The importance of having a reliable spam filter integrated with an email system cannot be understated. With new frauds and security threats popping up every day, business customers need to have a robust solution in place to protect against any attacks on their data or systems. Having a properly configured filter protects users from potentially harmful emails that can lead to stolen data or damaged infrastructure. This increases the overall level of security for the business while reducing costs associated resulting from successful attacks.

Spam filters are designed to identify and block messages that contain suspicious content by analyzing each message's origin, content, subject line, file attachments, and more - thus providing users with a comprehensive protection system against potential threats. Additionally, these filters often detect fraudulent websites or attempts at phishing frauds before they get through filters placed by firewalls or antivirus software; this ensures that users remain safe from potential malicious attacks even when browsing online or checking their emails on their mobile devices.

When selling spam filtering solutions, MSPs should emphasize how these solutions help businesses with protecting data as well as improving user productivity. In addition to blocking malicious messages from entering inboxes, filters ensure that employees are not wasting time sifting through irrelevant emails. This is done automatically by sorting out non-essential communications such as marketing emails or newsletters. This keeps workers focused on tasks at hand without being distracted by clutter in their inboxes. This allows businesses to save money by reducing wasted hours spent dealing with non-essential mail while increasing overall efficiency levels within the organization.

Overall, having an MSP provide integrative spam filtering solutions is essential for any business looking to stay ahead in today's competitive landscape. Not only does it protect against malicious threats, but it also improves user productivity levels while helping companies save money - making it an ideal choice for any organization looking for reliable ways of keeping their email system safe and secure.

**SPAM filtering products the MSP may use:**

- AVG Anti-Spam
- Mailwasher Pro
- McAfee SpamKiller
- Norton AntiSpam
- Cloudmark DesktopOne
- SPAMfighter Pro
- Kaspersky Anti-Spam
- Lavasoft Ad-Aware spam filter
- ZoneAlarm Spam Blocker
- Barracuda Spam & Virus Firewall

## Antivirus

Antivirus software is a critical tool for keeping computers safe from malicious threats such as viruses, trojans, worms, ransomware, and other malicious programs. Without adequate protection against these dangerous intrusions, businesses can suffer costly downtime and data loss. Managed service providers (MSPs) can provide this vital protection by centrally managing antivirus software to ensure clients' computers are properly protected.

MSPs provide robust antivirus solutions at an affordable price to their customers by utilizing cloud-based services requiring minimal setup and maintenance. This eliminates the need for expensive on-site hardware and infrastructure while still providing comprehensive security measures. MSPs can further reduce costs by utilizing automated tools for installing and updating antivirus software simultaneously on multiple systems enabling the quick deployment of security measures with minimal effort or disruption to existing infrastructures.

In addition to providing automated installation of antivirus software, MSPs can also centrally manage the software's configuration settings to make sure that the most up-to-date protocols and techniques are being used for protecting their clients' systems from infection. For instance, they can adjust settings such as real-time scanning frequency, type of scan performed (quick or full), level of heuristic scanning utilized (if any), and more. Thus ensuring maximum effectiveness against modern threats without compromising performance or user experience.

Furthermore, MSPs can also monitor virus activity across all managed systems to ensure that no infections have occurred. This allows them to quickly identify any areas where additional security measures may be needed to prevent the spread of malicious code or data loss due to ransomware attacks. As such, having an MSP provide a centralized security solution helps businesses stay one step ahead of attackers while reducing overall costs associated with responding to potential intrusions.

By providing a comprehensive antivirus solution through central management, MSPs help businesses protect their data from malicious threats while saving time and money, streamlining installation processes, and eliminating unnecessary hardware costs. Additionally, they help maintain maximum effectiveness against current threats through automated updates and configuration adjustments - ensuring clients remain safe from hackers both now and into the future.

**Some of the antivirus applications an MSP may use are:**

- Kaspersky Anti-Virus
- Norton AntiVirus
- McAfee VirusScan Plus
- AVG Antivirus
- Panda Security
- Bitdefender Antivirus
- Trend Micro Titanium Maximum Security
- Avast Free Antivirus
- ESET NOD32 Antivirus
- Avira Antivirus Personal

## Email Archiving

Email archiving is a crucial tool for any organization, particularly businesses and governments that work with sensitive information. Government entities must ensure their emails are securely backed up and stored securely to ensure compliance with regulations like the Freedom of Information Act (FOIA). For businesses working with the government, email archiving can help protect against potential legal issues, such as lawsuits or audit requests.

By properly archiving emails, businesses maintain an organized record of communication between themselves, their partners, and customers. This provides a solid evidence base should legal disputes arise in the future. Additionally, it makes it easier to search through historical emails to retrieve the relevant information needed for research or regulatory inquiries. With email archiving, organizations can have quick access to records from long ago that would otherwise be difficult or impossible to access.

Data security is another key factor for email archiving, especially when dealing with confidential information such as personally identifiable information (PII).

Creating a secure archive of all emails sent and received by employees within an organization allows companies to prevent unauthorized access to sensitive data and reduce the risk of data breaches. This helps organizations comply with data protection regulations like GDPR, HIPAA, and GLBA ensuring that personal data is adequately always secured.

Furthermore, email archiving enables companies to manage large volumes of messages more efficiently. Freeing valuable server space occupied by stored emails on local servers can be moved effectively to remote archives instead. With remote archiving organizations save costs associated with additional hardware upgrades while improving system performance thereby reducing strain on their networks. With automated backup processes in place, greater peace of mind is enjoyed knowing that critical business communications are securely preserved in the unfortunate event of accidents or disasters.

Overall, email archiving offers numerous benefits for businesses and government agencies alike. Archiving protects against potential legal issues arising from lack of evidence; provides better security for sensitive information; increases efficiency; and reduces costs associated with extra hardware upgrades due to large volumes of stored emails. Employers should strive to implement an appropriate solution for backing up and storing emails according to best practices, so they do not miss the various advantages provided in terms of productivity and data security.

**Email archiving applications an MSP may use:**

- Microsoft Exchange Server
- Mimecast Email Archiving
- G Suite Email Archiving
- EMC SourceOne
- Symantec Enterprise Vault

- HP Autonomy EAS
- Proofpoint Archive
- Cloudally Email Archiving
- ZixEmail Archive
- Red Oxygen Archive

## Backup

Backups are essential for the security of important data and information, as they enable organizations to restore operations in case of an unexpected disaster or other hiccups. An MSP helps businesses set up a system that provides them with both centralized and decentralized backups.

Centralized backups are typically done on a remote storage location, such as a cloud-based system or external hard drive. This type of backup is often preferred by organizations due to its convenience and cost-effectiveness as it allows them to store all their data in one place without having to maintain multiple physical storage devices. Additionally, centralized backups allow for automated processes that make it easy to monitor and update the system without manual intervention.

Decentralized backups involve storing data on multiple computers in various locations, whether they are located within the same building or at different sites around the world. This type of backup is beneficial when it comes to ensuring maximum security and reliability since each computer serves as an independent resource and so any problems affecting one device will not necessarily affect others. Decentralized backups can also be useful when there is a need for increased speed and efficiency due to their distributed nature. Different computers can be used simultaneously when performing tasks such as large downloads or uploads which would otherwise require a single server system.

An MSP helps businesses find the best solution when deciding between centralized or decentralized backups. They ensure that the organization's needs are met in terms of cost-effectiveness, reliability, performance, scalability, and security, while also recommending appropriate software solutions for backing up their systems based on specific requirements. Furthermore, MSPs also offer comprehensive support services for both types of backup systems; from setting up and testing before going live to providing ongoing maintenance and troubleshooting if necessary.

Having proper backup systems in place is more important now than ever before. Recent advancements in cloud technology make it easier than ever for companies to store their data remotely for future access if needed. An MSP will be able to provide advice and guidance when deciding between centralized or decentralized backups; helping organizations establish reliable recovery plans should something go wrong while staying mindful of budgetary constraints.

**Backup applications an MSP may use:**

- Symantec Backup Exec
- Veeam Backup and Replication
- IBM Tivoli Storage Manager
- Dell EMC Networker
- HP Data Protector
- CommVault Simpana
- Acronis Backup & Recovery
- Arcserve Unified Data Protection
- Veritas Netbackup
- StarWind Virtual SAN

## Cloud Solutions

When it comes to hosting and delivering servers, cloud computing is rapidly becoming the go-to for businesses of all sizes. The two main platforms that dominate the market in this regard are Microsoft Azure and Amazon Web Services (AWS). An MSP could use either one or both cloud solutions to provide customers with the best possible service.

Created by Microsoft, Azure is a cloud-computing platform which enables organizations to quickly and easily deploy, manage, and scale their applications without having to worry about hardware costs or maintenance. Azure provides an environment for users to develop diverse types of applications using multiple development frameworks, as well as an extensive suite of services such as analytics, networking, storage, and databases, making it ideal for enterprise-level architectures. Additionally, Azure offers advanced security features such as identity and access management which can help keep user data safe from unauthorized access or malicious activities.

On the other hand, Web Services provided by Amazon (AWS) is a comprehensive suite of cloud-computing that offers everything from storage and compute power to database systems, analytics tools, and application hosting. Much like Microsoft Azure, AWS provides users with a range of development tools ranging from DevOps automation tools to machine learning capabilities; allowing them to create powerful applications with minimal effort. Furthermore, AWS includes its own set of security measures such as encryption at rest/in transit and identity & access management, giving users peace of mind when it comes to protecting data in the cloud.

Leveraging either one, or both, of these leading cloud solutions, an MSP offers their customers fast and reliable server delivery services with minimum effort on their part. This allows businesses to focus more time on developing high-quality products instead of dealing with cumbersome hardware requirements often associated with traditional hosting models. Additionally, utilizing either platform ensures that customer data is kept secure and protected thanks to the advanced security features included in each solution. The result is to help businesses maintain compliance requirements while keeping operations safe from potential threats.

Overall, using either Microsoft Azure or Amazon Web Services can be beneficial for managed service providers wanting to offer efficient server delivery services without going through the hassle of managing physical hardware themselves. Taking advantage of these platforms' wide range of features allows the quick deployment of resources while maintaining control over how they are used. The result is deliverance of a better experience for customers while saving valuable time and money in the process.

## Financial Packages

Managed Service Providers (MSPs) can be a great asset to businesses wishing to support financial applications such as QuickBooks, Sage, Peach Tree, among other financial applications. MSPs provide businesses with an array of services that can help simplify the process of deploying, managing, and maintaining these applications.

For instance, MSPs can provide businesses with expertise in setting up and configuring their software for optimal performance on their hardware. This includes ensuring that all required software components are installed correctly and that necessary dependencies are satisfied. Additionally, MSPs can assist in implementing proper security measures such as encryption and authentication protocols to ensure the safety and protection of sensitive data from unauthorized access or malicious activities.

Additionally, MSPs can monitor the system's performance continuously aiding in the identification of potential problems before they affect operations or cause downtime. Furthermore, an MSP may offer additional services designed to help maintain the overall health of the system hosting these financial applications. This includes regular updates and patches to address any new vulnerabilities or issues that may arise over time, keeping them running smoothly without requiring manual intervention.

In addition to providing technical support for financial applications, MSPs assist businesses in developing effective strategies for handling finances by offering tailored advice based on their specific needs. This could involve suggesting diverse types of investments or creating budget plans to maximize profits while minimizing risk. Additionally, an MSP could provide guidance regarding tax planning as well as other helpful tips related to efficient accounting practices, thus helping businesses stay on top of finances with minimal effort.

Overall, MSPs can be extremely beneficial for businesses looking for ways to get more out of their financial applications without having to take on all the associated tasks themselves. Leveraging an MSP's expertise in both technology and finance allows businesses to focus more time on expanding their operations, leaving behind the worry over complex server configurations or tedious bookkeeping tasks. This leads to higher efficiency in terms of both costs and resources, which allows organizations of all sizes to reap the benefits of cost-effective yet reliable financial solutions without sacrificing quality or security in the process.

**Finance software applications an MSP may support:**

- QuickBooks
- Sage

- Peach Tree
- Xero
- Microsoft Dynamics GP
- Intuit Financial Management Solutions (IFMS)
- Quicken Home & Business
- FreshBooks Cloud Accounting Software
- Wave Accounting Software
- Zoho Books

## Adobe Product Line

Managed Service Providers (MSPs) are an asset to businesses wanting to boost productivity or achieve specialized functions and workflow with the Adobe product line. Leveraging an MSP's expertise in technology and software allows businesses streamlining of processes while minimizing time and money spent on manually configuring and maintaining their systems.

For instance, an MSP can provide support when deploying or updating Adobe Creative Cloud applications such as Photoshop, Illustrator, Premiere Pro, and InDesign. This includes ensuring that required components are installed correctly and that necessary dependencies are satisfied before launch. Additionally, they assist with integrating other tools into the Adobe ecosystem for expanded functionality through APIs or plugins.

Moreover, an MSP can help businesses ensure systems remain secure when running Adobe products. An MSP could implement proper authentication protocols such as two-factor verification or digital signatures to protect against malicious activities, keeping sensitive data safe from unauthorized access or manipulation. Furthermore, they could offer advice on best practices for patching vulnerabilities and regularly updating software, thus ensuring user machines remain up to date with the latest security features released by Adobe.

In addition to technical support for Adobe applications, MSPs could provide expert guidance on how to maximize their use within certain industries or business areas. This includes offering advice on how to effectively manage workflows between different teams, establishing efficient communication channels between clients and designers, or optimizing team collaboration through Project Felix's 3D compositing capabilities. Additionally, they could suggest methods of using Adobe tools such as Stock Photos for improved marketing campaigns, in turn helping businesses better reach their target audiences with engaging website of social media content.

Ultimately, working with an MSP can be highly beneficial if a business wants to attain more from the investment in the Adobe product line without worrying about managing its IT infrastructure. Taking advantage of this platform's wide range of features enables the deployment of resources while maintaining control over usage leading to a better experience for their customers while saving valuable time and money in the process.

**Adobe products an MSP may support:**

- Adobe Creative Cloud
- Photoshop
- Illustrator
- Premiere Pro
- InDesign
- Stock Photos
- Project Felix 3D Compositing
- Acrobat Reader DC
- Adobe XD
- Adobe Dreamweaver

# Custom Applications

Managed Service Providers (MSPs) can be incredibly useful for businesses needing to deploy in-house or third-party custom applications. By leveraging an MSP's expertise in technology, businesses can take advantage of the best practices and experience necessary to ensure their applications are running securely and efficiently.

When it comes to developing custom applications, an MSP can provide valuable support throughout the entire process – from software design to deployment. During the design phase, an MSP analyzes existing systems for potential compatibility issues as well as identifying areas where new features might be beneficial. They can make suggestions regarding technologies that are best suited for the application, such as choosing a cloud platform like Amazon Web Services or Microsoft Azure for hosting. Upon completion of development, an MSP can assist in deploying the application across multiple servers over a secure network, while ensuring the system is properly configured for optimal performance.

In addition to software design and deployment, MSPs can help with integrating additional products into existing systems. This includes providing advice on how to use specific APIs or plugins to link external applications together, as well as suggesting software tools to streamline workflows between different departments within the business. Furthermore, they could offer insight on how best to utilize these tools when creating customized experiences tailored specifically for customers, making sure all key features are present while maintaining a user-friendly interface.

Finally, an MSP could be instrumental in providing ongoing maintenance and support services once a custom application has been deployed. This includes patching any security vulnerabilities discovered during regular scans, performing backups of data stored within the application, and installing regular updates that add new features or fix bugs found during testing phases. This ensures that any investment made in developing custom applications is continually paying dividends, keeping them running smoothly and securely, all the while performed on time with minimal disruption. All these advantages make an MSP a highly valuable option for any business looking for reliable IT solutions.

# Steps to Implement and Maintain Managed Services

## Define Your Needs

It is essential to thoroughly define your company's needs before selecting a managed services provider. Knowing exactly what your business needs from the provider and having a clear understanding of how those needs will be addressed is key to selecting the right fit for your organization. A thorough assessment of your company's current IT infrastructure, as well as its future growth plans, will ensure that the managed services provider can anticipate and meet any changing demands.

Having a firm grasp on what type of resources and support you need to maintain your desired operations and performance levels is also important. This includes factors such as software compatibility, hardware maintenance, data storage requirements, security protocols, backup strategies, scalability options, customer service levels, problem resolution times, and more. With this information in hand, it becomes easier to identify which service providers are best suited to provide the necessary support.

Defining your company's needs also helps guide when it is time to upgrade existing systems or implement new ones. For example, if certain operations are taking too long due to outdated systems or software issues, then it might be time for an upgrade or update with the help of a managed services provider. Similarly, if your business is expecting rapid growth or experiencing rapid changes in technology trends then it could be beneficial to acquire additional help from a managed services provider.

In conclusion, defining your company's needs before engaging with a managed services provider is critical for making an informed decision on who can best serve you now and in the future. It enables you to take stock of all required resources and adequately plan how they will be serviced going forward so that no stone is left unturned when choosing the right managed services provider for your business.

## Discuss Implementation Details

Before signing a contract, make sure to discuss the details of services and associated costs. This includes service level agreements, fees for additional services or maintenance, communication expectations, and problem resolution times to name some of the more common contract details.

It is critical to set up monitoring: Establishing reliable performance baselines is essential to ensure that you can quickly identify problems as they arise. With a monitoring system in place, you can track performance metrics such as network speed and uptime, storage space usage, system load times, and more. This data will help the managed services provider identify any potential issues early on and provide timely solutions to keep your systems running at peak efficiency.

It is of utmost importance to understand the details of the services and associated costs before signing a contract with an MSP. As previously mentioned, this includes having a thorough understanding of service level agreements, fees for additional services or maintenance, communication expectations, and problem resolution times. Understanding these details ensures that your business is not only getting the best possible support and services but that you are also aware of any potential hidden costs or issues that may arise during implementation. This knowledge allows an informed decision when selecting the best MSP for your business needs and expectations.

A clear understanding of the implementation details can also help you stay ahead of any issues or problems that may arise in the future. With regular monitoring and tracking in place, you can quickly identify any irregularities in performance metrics such as network speed and up-time, storage space usage, system load times and such, which allows your provider to provide timely solutions to keep systems running optimally. Additionally, having a reliable performance baseline in place helps ensure that any changes made by the provider meet their intended purpose rather than creating new problems.

Overall, it is essential to have an in-depth understanding of all services and associated costs before agreeing with a managed services provider. This knowledge will give you peace of mind knowing that the provider you select has met all your requirements and is equipped to deliver quality service on time and within budget. Furthermore, being able to track system performance over time will ensure that any changes made improve operations rather than creating new issues for your IT infrastructure.

Choose wisely: Make sure that you get a good sense of the managed services provider's culture, capabilities, customer service levels, and problem resolution times before signing on the dotted line. Taking time to do your research helps ensure that you are making an informed decision best suited for your business overall. Doing this will pay dividends in the future as you move forward with a managed services provider that can meet the requirements of your organization.

# Research Providers

Use online searches to find a reputable provider to meet your business needs. Make sure to read reviews and get customer feedback on each option.

When selecting a provider for computer-managed services, it is important to contact multiple providers to assess their offerings and determine which one best meets your organization's needs. When speaking with potential providers, make sure to ask questions about features and pricing structures so that you can compare the different options and make an informed decision. It is also beneficial to inquire about their implementation timelines and maintenance plans - understanding these details will ensure that the service is implemented correctly and operated reliably. By taking the time to carefully select a provider and understand their services, organizations can ensure they have the right solution in place to meet their needs

# Draft a Contract

Once you have finalized your decision on which managed services provider best suits your needs, it is time to draft a service level agreement (SLA). This should outline any deliverables from the provider as well as terms for payment and termination. It is also important to include details about how any issues or conflicts will be managed, and how performance metrics are measured. Having a written contract in place is key to ensuring that both parties understand their obligations and can work together successfully over the long term.

Following these steps ensures that your business receives the best services from an MSP and stays up to date with all its technological needs. With the right provider, you can rest assured knowing that your systems are secure and running smoothly – allowing you to focus on building your business.

## Contents of an MSP-Client Contract

A managed services contract is defined as an agreement between a managed services provider (MSP) and a client. The contract should outline the scope of services that the MSP will provide, service-level agreements (SLAs), payment terms, and any exclusions or limitations.

Service Hours: The MSP must provide services during the agreed-upon service hours. This should include an estimated response time to requests for assistance, as well as resolution times for issues that may arise.

Scope of Services: All services provided by the MSP must be detailed in the managed service agreement, including routine maintenance and any exclusions from these services.

Service-Level Agreement: The SLA outlines expectations for response times and resolution times for issues that may arise. It should address customer satisfaction guarantees and acceptable levels of performance.

Payment Terms: The payment terms should clearly define when payments are due from the client and what methods of payment are accepted by the MSP, such as credit card or electronic payment transfer (EFT).

Exclusions/Limitations: Any exclusions or limitations to services provided by the MSP should be outlined in detail to eliminate confusion over coverage.

Taking all these considerations into account when crafting an MSP contract enables both parties to rest assured of having a clear understanding of their respective rights and obligations under the agreement.

## Implement the Services

After signing the SLA, work with the provider to implement the necessary managed services. During this process, make sure all stakeholders are kept in the loop about progress and any changes or delays. Ensure that the provider is meeting all obligations and taking full responsibility for issues that arise. At the end of implementation, review the services to ensure they are fully functioning as expected.

It is important to remember that working with an MSP requires ongoing collaboration and communication between both parties. Be sure to keep up regular communications and take immediate steps whenever problems arise. By doing this, your organization can maintain its technology systems efficiently, securely, and cost-effectively. Selecting a managed service provider is not a decision into which you should rush. It is essential to do the research, speak with potential providers, and make an informed decision based on the needs of your business.

## Monitor Performance

Once implemented, maintain regular communication with the provider and monitor the performance of their services over time. Address any issues immediately to ensure satisfaction for all parties involved.

Most providers will avoid providing voluntary metrics and key indicators. Make sure your contact states which indicators you should receive in the form of a report and how often.

## SLA (Service Level Agreement)

A service Level Agreement (SLA) is a legally binding agreement between a service provider and a customer, which outlines the terms of services for the customer. It is typically used in the context of IT services and should include information such as availability levels, response times, cost structures, and resolution times. SLAs are put into place to ensure that both parties are held accountable for any agreements made and that the customer receives the agreed upon quality of service. By setting clear expectations up-front, businesses can ensure their managed services provider is delivering what has been promised and hold them accountable in the event of any discrepancies.

SLAs should be regularly monitored to ensure they are being met and should be updated as needed to reflect changing business needs. If necessary, it is important to renegotiate the terms of the agreement to ensure customer satisfaction. By consistently tracking service performance against the SLA, organizations ensure they are utilizing their managed services provider effectively and receiving full value for their money.

Overall, it is important to carefully select a managed services provider and understand their services. When speaking with potential providers, ask questions about features and pricing structures. As previously mentioned, draft an SLA that outlines expectations for both parties and monitor performance against this agreement regularly to ensure satisfaction for all involved. By doing so, businesses can have peace of mind that their managed services provider is meeting their needs and delivering the quality of service that was promised.

The SLA must include basic metrics. Response and resolution time are most important. Some service providers may become resourceful by producing first response, progress reporting, and resolution times as separate factors for more freedom in satisfying customer needs. All these diverse types of responses have been designed to give the provider better tools for meeting customers' demands. Nevertheless, it is resolution time that typically matters most to customers.

It is necessary to understand the response time for any issue needing attention, and it is beneficial to get a monthly report with data from all work orders. Never accept quarterly, semi-annual, or annual reports with the SLA. These types of reports do not allow an opportunity to take corrective action when problems arise. Immediate access to details and metrics do allow quick action and review. When the yearly review is initiated, annual reports should be submitted in conjunction with the monthly report.

## Key Indicators and Metrics

When choosing an MSP, it is important to ask for key indicators and metrics. This enables the customer to gauge whether they are receiving the quality of service agreed upon in the SLA. Some key indicators that should be looked at include availability levels, response times, cost structures, and resolution times. It is beneficial to investigate other factors such as monthly reports with data from all work orders to gain better visibility into the service performance. Additionally, businesses should make sure that their SLA includes clear expectations for both parties and ensures that details on response time and resolution time are included.

It cannot be overstated that quarterly or even semi-annual reports on SLA should be avoided as these reports will not provide businesses with adequate visibility into any issues that may arise. Instead, companies should insist on monthly reports to get a more comprehensive view of performance against an agreed-upon SLA. In addition to monthly reports, yearly reviews can be conducted along with an annual report which provides a summary outlining any changes or areas where improvement has been made over the year. Businesses should ensure that their provider outlines how often they will receive reports and which indicators can be expected in these forms. This ensures that their managed services provider is meeting their needs and delivering the expected quality of service. Organizations should make sure they receive timely non-conformance reports to quickly address problems that arise. This can be done by requesting regular status updates and having a clear communication plan for several types of notifications.

Understanding system expiration, renewals, and warranty information is essential for any organization using managed services. With an understanding of these vital details, organizations ensure that their systems are running optimally and that they are getting the most value from their managed service provider. It is also important to be aware of potential expiration dates or warranties before they become obsolete so that organizations know when it may be necessary to upgrade or replace a system. Additionally, awareness of system expiration, renewals, and warranties allows organizations to plan and budget accordingly for needed future upgrades or replacements. Knowing this information maximizes efficiency while minimizing costs.

# Resolving Issues with Managed IT Services

## Understanding System Expiration, Renewals, and Warranties

Keeping track of system expiration, renewals, and warranties is an essential part of using managed IT services. By understanding these details, organizations can ensure that systems are running optimally, obsolete systems are avoided, and the most value from their service provider is obtained.

## Requesting Regular Status Updates and Establishing a Communication Plan for Notifications

Requesting regular status updates and establishing a communication plan for notifications are essential steps for businesses utilizing managed services. Companies should ensure that their provider outlines how often they will receive these updates.

It is also important for businesses to establish a clear communication plan for types of notifications. This ensures that any technical issues or disruptions are quickly addressed as soon as they arise. Organizations should set up different notification methods such as email, text messages, or social media alerts to ensure that notification is received promptly regardless of the situation. Additionally, businesses should consider setting up automated alerts which can be sent when certain performance metrics fail to meet the agreed-upon SLA standards. This allows organizations to stay informed and respond sooner rather than later to any potential issues that may arise with their managed services provider's performance.

Requesting regular status updates and establishing a communication plan for notifications gives businesses greater visibility into how their managed services provider is performing and allows them to address any issues quickly. Taking proactive steps such as these enables organizations to have surety that they are receiving the expected quality of service.

## Utilizing Remote Access Solutions for Immediate Support When Needed

Utilizing remote access solutions for immediate support when needed is an optimal method for businesses to ensure their managed services providers are meeting the agreed-upon service level agreement (SLA). Remote access solutions provide organizations with the ability to quickly troubleshoot and address any issues without having to wait for onsite assistance. This is especially beneficial in situations where critical systems are down or experiencing performance not meeting expected levels.

When utilizing remote access solutions, organizations should make sure they have full visibility into what is happening with their systems so they can effectively troubleshoot any technical problems. Establishing secure protocols and authentication methods is important to ensure that only authorized personnel have access to sensitive information and data. Thus, having these processes in place, organizations can rest assured knowing their data and systems are secure from potential threats or malicious actors.

It is essential for businesses to ensure they have the

right tools and personnel available for timely support when needed. Organizations should consider investing in quality monitoring tools and establishing resolute teams of IT technicians who can respond quickly to any incidents or service requests – regardless of time of day. Companies should evaluate their SLA metrics regularly, so they know they are getting the most out of their managed service provider.

In summary, utilizing remote access solutions allows businesses to quickly diagnose and address technical issues immediately while increasing overall system security. Having dedicated teams of experienced IT professionals as well as reliable monitoring tools allow businesses to receive the most from their managed services provider while providing peace of mind knowing that all aspects of system performance are monitored closely. With these measures in place, organizations can confidently trust their provider to deliver high-quality service that meets all agreed-upon expectations outlined in the SLA.

## Invest in Cybersecurity Measures to Protect Your Network Against External Threats

Investing in robust cybersecurity measures is essential for organizations looking to protect their network from external threats. Cybersecurity measures ensure that critical data and systems remain secure and protected from malicious actors, ransom ware attacks, and other security threats. Businesses should consider investing in a multi-layered approach to cybersecurity to include implementing next-generation technologies such as advanced firewalls, user authentication methods, Intrusion Prevention Systems (IPS), encryption solutions, identity management solutions, and malware scanners.

Advanced firewalls provide organizations with an additional layer of protection allowing only authorized personnel access to networks. Firewalls can be configured to block certain types of traffic and restrict access to specific network services or websites based on rules set by the administrator. User authentication methods such as two-factor authentication or biometric identification provide an extra layer of security when accessing sensitive information or data.

Intrusion Prevention Systems (IPS) use signatures or heuristics to detect malicious activity such as suspicious traffic patterns or potentially dangerous malware payloads on a network. IPS can be used to monitor malicious activity in real-time and alert administrators, when possible, threats are detected. Encryption solutions can be used to protect sensitive data as it is transmitted across a network ensuring only authorized users have access to this information.

Identity management solutions ensure secure authentication across multiple systems using strong passwords and multi-factor authentication methods such as biometrics or one-time passwords (OTPs). These tools can be used to provide organizations with detailed logs of who has accessed their systems and help to track user behavior while reducing the chances of unauthorized access. In addition, organizations should consider investing in reliable malware scanning solutions. Malware scanners look for potential threats regularly and alert admins if any suspicious files are detected on the network.

In summary, investing in a comprehensive cybersecurity strategy provides organizations with better visibility into how their systems are being used while providing greater control over who has access to critical information and data. Taking measures such as these assures that networks remain secure against external threats without sacrificing performance or user experience. With the right tools in place, businesses not only improve the overall security posture of their organization but also establish greater trust among customers through the provision of protecting sensitive data.

## Keeping Track of Changes in Technology

Keeping up with changes in technology is critical for businesses to stay competitive and ahead of the curve. Companies need to be able to adapt quickly and understand the implications of any recent technologies affecting their operations. Organizations must devise a plan that will allow them to quickly assess, analyze, and respond to any changes that may occur.

One way for businesses to monitor changing technologies is to stay informed about industry trends and review emerging technologies that could potentially benefit their organization. Companies should take time to research new products or services that can streamline their processes while looking into how they can leverage existing solutions and systems to improve efficiency. In addition, staying abreast of news related to cyber security threats helps businesses to prepare and prevent malicious attacks on their network.

Training programs are another way companies can keep track of changing technologies. Investing in these programs for their employees provides a better understanding of how new tools or processes work and how they can be used most effectively. This ensures that all members of the organization are up to date with the latest techniques, skills, and knowledge needed to remain competitive. Furthermore, having a centralized system for tracking changes and updates within an organization's IT infrastructure allows employees across departments to access the same information and communicate more efficiently when responding to rapidly changing circumstances or environments.

Companies need to evaluate their current practices regularly so they can identify areas for improvements, or adopt modernization quickly without sacrificing performance or efficiency. The use of analytic-driven decision-making allows organizations to measure the impact of certain strategies while providing insight into what works best within an organization's specific context. This helps businesses identify opportunities where they can adopt relevant technological solutions without disrupting existing processes or operations. Managed services assist when it comes to managing multiple systems, software upgrades, data recovery, server maintenance, cloud migration, data security, and more which makes them an invaluable asset for businesses looking to stay up to date with technological advancements. Overall, keeping track of ever-changing technologies requires organizations to have a proactive approach towards monitoring shifts in the industry landscape as well as investing in employee training programs and updating IT infrastructure frequently.

Overall, staying on top of ever-changing technologies requires organizations to have a proactive approach towards monitoring shifts in the industry landscape as well as frequently investing in employee training programs and updating IT infrastructure. Doing this, businesses will not only increase productivity levels but create a culture where employees feel empowered when faced with difficulties or complexities due to technological advancements. Taking steps such as these ensures that organizations remain competitive no matter the environment and remain successful during times of rapid transformation.

# Advanced Concepts

At times, businesses require solutions that go beyond the typical practices. For instance, an MSP could be enthusiastic about offering help with emailing services, web hosting support, onsite server management, or cloud server administration.

To help your business expand its reach and increase its worth to clients, you may need solutions not offered by all Managed Service Providers. Consider potential options for taking your operations to the next level.

## The Client Portal

A client portal can be a powerful tool for businesses to improve customer service and provide an organized, streamlined experience. A client portal allows customers to access information concerning orders, contact customer service representatives and view any available resources such as product tutorials or frequently asked questions. By providing easy access to all the relevant information in one place, the client portal eliminates the need for customers to search through pages of text or documents for specific answers. Furthermore, as inquiries are managed more efficiently it helps reduce customer wait times.

Client portals can also help businesses stay organized, providing insight into customer needs from a single platform. All customer-related data can be stored in one centralized location, making it easier for employees to find critical information quickly when responding to inquiries or resolving issues. This helps reduce staff workloads significantly and enables businesses to streamline operations with improved efficiency. Additionally, customer profiles can be created so that each customer's preferences and previous interactions with the company are saved. This allows companies to anticipate customers' needs before they even make contact and deliver tailored experiences accordingly.

Client portals provide enhanced security measures that protect sensitive data, which is especially beneficial when dealing with financial institutions or other organizations overseeing copious quantities of confidential information. Advanced encryption protocols ensure that customer data always remains secure. Encryption protocols provide additional protection against malicious attacks by identifying suspicious activity quickly and blocking unauthorized access attempts immediately.

In summary, a client portal can be beneficial for businesses looking to improve their organizational capabilities and maintain prominent levels of customer service. It provides customers with a convenient way to access information about their accounts while arming companies with greater insight into their target audience's preferences so that they can deliver personalized experiences. In addition, it enhances security measures keeping sensitive data safe from cyber threats which creates a higher level of trust among customers knowing that their data is always well-protected. Therefore, implementing a robust client portal is essential for businesses looking to stay organized in this digital age and remain competitive in the modern marketplace.

# The LMS (Learning Management System)

An effective Learning Management System (LMS) can be an invaluable asset for businesses looking to streamline operations and improve customer service. An LMS is a software platform that provides an integrated suite of tools for training delivery, course management, analytics, reporting, and other administrative tasks. These tools give businesses the ability to create and manage online courses efficiently, track learner progress, and provide feedback on performance quickly.

The benefits of implementing an LMS are numerous. For starters, an LMS makes delivering training content more accessible to customers since they can access the material from any device at any time. This eliminates the need to attend physical classes or seminars while cutting down on costs such as travel expenses or physical materials associated with hosting events. Furthermore, learners can take courses at their own pace as they have full control over the learning speed and timeline which boosts engagement levels significantly.

An LMS provides businesses with powerful insights into customer needs by tracking learner progress. Closely monitoring user activity in real-time, businesses gain valuable insight into topics that resonate best with different audiences allowing work to improve in those areas. Furthermore, all data is stored in one centralized location which allows companies to have a single view of customers from a sole source of truth. This makes it easier to identify patterns quickly and make informed decisions about training strategies.

Moreover, an LMS enables companies to quickly assess performance and provide feedback as needed throughout a course's duration without having to manually grade assessments or exams. This reduces staff workloads significantly while providing learners with timely feedback that will help improve their understanding of the material faster than traditional methods would allow. Additionally, an LMS makes it easy for companies to automate evaluations by setting up rules so that assessments are graded automatically giving employees more time as well as resources to focus on more pressing tasks such as designing new courses or creating engaging activities for learners.

In summary, a Learning Management System offers numerous advantages to help businesses streamline operations and improve customer service while reducing costs. With the array of features in an LMS, including custom content delivery platforms, automated assessment grading, real-time data analytics, and powerful reporting capabilities - businesses easily deliver personalized training experiences. The training can be tailored specifically towards learner needs while gathering meaningful insights into customer preferences along the way. Implementing an effective LMS is essential for staying competitive in today's marketplace. This is due to its immense potential for improving customer engagement levels and optimizing business operations quickly and cost-effectively.

## Document Management System

A Document Management System (DMS) is an effective system for storing and managing digital documents. A DMS allows businesses to organize data into a single repository. With DMS, businesses can easily store, access, and share documents from anywhere at any time, eliminating the need for physical storage or manual labor. This type of system ensures document security through controlled access, making it easier to protect confidential information against unauthorized users and hackers.

One of the key benefits of a DMS is the simplification of the process of searching and retrieving documents. Using advanced search capabilities such as tagging, keywords, and metadata, users quickly locate files within a fraction of a second. This saves an immense amount of time compared to traditional methods of manually searching folders or cabinets. This system allows multiple users to collaborate on documents simultaneously while tracking all changes made by each user in real time. This prevents overriding another user's changes.

DMS eliminates the need for paper copies as documents can be accessed digitally with just an internet connection, reducing clutter and costs associated with printing materials. Additionally, because data is stored in one centralized location - everyone has access to the same version of the document, reducing errors caused by out-of-date copies being disseminated throughout the organization.

Another advantage of DMS is the ability to automate routine tasks such as version control or archiving processes, reducing staff workload significantly while providing more accurate results than manual processes would allow. This type of system provides businesses with enhanced security features such as user authentication measures such as passwords or biometrics keeping sensitive data safe from malicious intruders. Comprehensive audit trails make it easy to track activities within the system thus helping organizations stay compliant with regulations such as GDPR or HIPAA without having to manually monitor user activity.

Overall, a Document Management System offers numerous advantages helping businesses streamline operations while saving time and money. Connecting teams regardless of geographical location, reducing storage costs, automating tedious tasks, providing enhanced security features – allows companies more efficient workflows with fewer resources. This frees staff members to pursue other pressing tasks. The cost savings associated with using a DMS are immense when done correctly, these savings range from labor savings due to automation processes, reduced storage costs due to digital records, improved accuracy due to lone source repositories, and enhanced security measures due to user authentication options. A DMS is essential for staying competitive in today's modern marketplace while ensuring customer service levels remain high throughout every interaction with your business. As an added advantage to DMS there is improved customer experience

## Website

When it comes to website development, many MSPs require the business to hire a separate firm. To stay within budget and get reliable results, look for website-managed service providers. Quality companies can offer excellent value and excellent customer service too.

Having a website that is clean, fully functional, and up to date is critical for business in today's digital world. A business's website serves as the primary point of contact with potential customers and must be able to provide accurate information quickly and efficiently. A website should be easy to navigate with clear menus, categories, and pages enabling visitors to find what they are looking for quickly without becoming overwhelmed or frustrated. The website should have a professional design that reflects the company's brand and image while being up to date with current trends in web design.

The content of a website should be well-written, informative, and engaging for readers. This entails not only providing visitors with basic information about the company but also detailed information about products/ services offered, customer testimonials, news & events related to the company, and other relevant topics. The content should not just be informative; it must be persuasive and contain keywords targeted toward potential customers. This will increase search engine optimization (SEO) rankings which in turn lead to increased customer visibility on search engines such as Google or Bing.

In addition to having clean design elements and relevant content, a website must have proper back-end functionality to ensure it always runs smoothly. Everything from hosting services to coding languages should be up to date allowing features such as shopping carts to work properly when needed. Of further importance is the assurance that databases can store copious amounts of customer data securely without crashing due to server overloads. Websites must remain secure both externally (from hackers) via firewalls or encryption systems as well as internally (from internal employees) by utilizing user authentication methods such as passwords or biometrics for login access control purposes.

Overall, a clean, fully functional website is vital for businesses in today's digital age. It allows companies to create an online presence that accurately reflects their brand while providing visitors with reliable information quickly and efficiently through an intuitive design interface that remains secure regardless of who visits or how often they do so. A well-developed website is essential for staying competitive in today's global market while ensuring customer service levels and experiences remain high throughout every online interaction with your business.

# WordPress for Websites

WordPress websites are prevalent and have remained a top choice for many reasons. Choosing to work with WordPress undoubtedly reaps a wide range of rewards.

WordPress has become the dominant website platform for several reasons. There are an array of features and benefits, such as cost savings and easy-to-use content management systems, making it extremely attractive for businesses looking to launch or expand their web presence.

For starters, WordPress is free and open-source software, meaning businesses avoid spending thousands of dollars on proprietary website platforms. Additionally, its intuitive user interface provides ease of use for anyone with minimal experience to create beautiful websites with powerful features. WordPress offers extensive customization options, allowing users to tailor websites to their branding needs without relying on expensive professional designers or developers. Furthermore, WordPress sites are secure and regularly updated with the latest security patches - making them safer than similar platforms not receiving regular maintenance updates.

Additionally, WordPress provides access to thousands of additional plugins and themes (some free and some paid) allowing adaptation to any requirement imaginable. For example, numerous e-commerce plugins are available for the platform enabling businesses to easily add a shopping cart feature to their website - eliminating the need for costly custom development work. Similarly, there are countless SEO plugins available that can optimize a site's ranking in search engine results pages (SERPs). This helps companies increase their online visibility and attract potential customers - resulting in increased sales and higher profits over time.

Finally, because WordPress is so popular among users, there is an abundance of online resources available for those wishing to learn more about using the platform and customizations. From tutorials and forums, to dedicated support teams from available hosting providers. Compared to other website platforms, finding answers is easy when working with WordPress.

Businesses looking for a cost-effective way of creating a professional website have found tremendous success by leveraging the power of WordPress as their web development platform of choice. Not only does WordPress require minimal financial investment; but its vast array of features makes it possible for businesses large or small to create stunning websites. All the advantages that come with robust security and reliable maintenance updates are present with WordPress leading many companies toward increased financial gains where they least expect it.

## Appealing MSPs not Providing Support for All Services

Firstly, Managed Service Providers (MSP) typically present various plans to tackle the company's IT needs and configurations. The provider can manage third-party vendors on behalf of the business if you decide that this will be their responsibility. An MSP can take care of all chosen network management solutions as well. When considering signing a contract with an MSP, do not hesitate to ask if they will manage other vendors on your behalf. In most instances, for only a few extra dollars, you can request that the provider takes control of all entities and manages any additional managed services providers as needed. Do not miss this opportunity to secure peace of mind and get maximum value from your agreement.

# MSP vs MSSP

Managed Service Providers (MSPs) and Managed Security Service Providers (MSSPs) can both provide managed services to their customer base. An MSP offers a variety of IT-related services, such as network setup, device management, software installations, support services, and more. An MSSP is a specialized form of an MSP focusing on providing security services for its clients. This includes areas such as antivirus protection, malware scanning, firewalls, and other similar solutions.

When it comes to the differences between an MSP and an MSSP, the biggest one is that the latter focuses on security-related services while the former provides multiple IT-related solutions. Additionally, MSSPs have specialized tools and staff members experienced in information security. While an MSP may be able to provide basic security solutions like antivirus software or malware protection with its existing resources, these must be tailored to fit each client's unique needs to be effective. On the other hand, an MSSP will typically have more advanced technology to offer which allows them to customize their service offering according to the specific requirements of each customer.

Furthermore, another difference between an MSP and MSSP lies in the scope of their service delivery capabilities. Both types of providers typically offer ongoing support contracts with a certain degree of customization options available based on individual customer needs. When it comes to updates or upgrades required due to technological advancements, or changes in regulations or compliance standards, this task usually falls under the purview of an MSSP partner. This is due to greater knowledge about data security protocols. As a result, this makes an MSSP better suited for clients addressing ever-changing cyber threats and staying ahead of the game when it comes to cyber defense.

MSP (Managed Service Provider) and MSSP (Managed Security Service Provider) provide distinct functions for different customers. MSPs offer a broad range of services helping customers manage technology, including cloud computing, network services, storage, data centers, desktop support, and software management. They also provide coordination and integration of these services to ensure all IT systems work optimally together. In contrast, MSSPs focus specifically on cybersecurity, providing customers with the necessary security tools to protect data and networks. MSSPs are typically responsible for monitoring customer networks for potential threats and vulnerabilities, responding appropriately to any identified threats or security incidents, configuring firewalls and other security measures to protect customer systems from malicious activity. This includes regularly scanning networks for emerging threats. An MSSP may also provide additional services such as employee training on cybersecurity best practices or malware protection strategies. Outsourcing the responsibility of protecting against cyber threats to an experienced MSSP partner, ensures businesses that their data is safe from malicious attacks while freeing up internal resources for other important tasks.

## Quality Assessment

When it comes to quality assessments, there are some key differences between an MSP and MSSP. An MSP provides services that fall into the category of "keeping the lights on" -meaning they maintain basic IT infrastructure and assist with tasks such as software installations, device management, and network setup. The MSSP, on the other hand, focuses more heavily on security-related services. This includes activities such as malware scanning, firewalls, antivirus protection, and ongoing monitoring of customer networks for potential threats.

In terms of quality assessment specifically, an MSP will provide regular maintenance checks to ensure systems remain up to date with any changes in technology or regulations. MSPs help customers develop best practices for system use and support that aligns with internal goals and objectives for a successful workplace IT environment. An MSSP focuses on providing advanced security measures tailored to each client's unique needs to protect data from malicious activity. This might include advanced monitoring capabilities, automated alerting protocols in response to suspicious activity or changes in customer networks, patching software regularly with latest security updates, or offering employee training courses on cybersecurity best practices.

In summary, both types of managed service providers are helpful when it comes to ensuring quality assessments within a business environment; however, their approaches differ due to the distinct services they offer. Understanding how these two providers differ when it comes to assessing quality enables organizations to make informed decisions about which type of managed service provider suits their needs.

## MSP & MSSP Together

Typically, MSP and MSSP are separate entities; however, there are instances when managed service providers also offer security services. This combination can be beneficial for businesses requiring comprehensive IT and cybersecurity solutions. Having both services under one roof, organizations save time and eliminate redundant tasks while coordinating the operations of different departments. Furthermore, using a combined service provides the added benefit of having a single point of contact for all IT needs as well as access to additional resources.

The greatest advantage to working with an MSP/MSSP hybrid is the extra security provided by combined services. Having a single provider experienced in providing both technology management services and cyber security strategies enables customers to protect data from malicious attacks while managing their IT infrastructure efficiently. Additionally, many MSPs that offer MSSP capabilities have state-of-the-art tools and expertise needed to detect potential threats before they cause damage or disruption to a company's operations. An MSP/MSSP hybrid eliminates the need for customers to hire additional employees or train existing ones on best practices for cyber security protocols as this is overseen by the provider.

On the downside, depending on the scope of work required and the specific needs of a customer's network environment, hiring a single provider for both services may not always be cost-effective. It is important to research and compare providers thoroughly before committing to contracts. This ensures that they are offering competitive pricing and quality service. Additionally, customers should understand what types of services are included in any package purchased to avoid invoicing or support requests surprises down the road.

# Equipment Purchases

## Questions on Purchasing

The purchase of necessary equipment, such as computers and software, is an important part of any business. The purchase process should be planned to ensure that appropriate items are bought at a reasonable cost. Research should include education on various products, comparing prices and features, understanding warranty details, and ensuring compatibility with existing systems. It is important to budget for training costs to maximize the use of purchased equipment.

Finally, every purchase should have a plan for disposal or recycling when it reaches the end of its useful life. This minimizes waste and keeps costs down, avoiding unnecessary replacement purchases. Managing equipment inventory helps reduce losses due to theft or misplacement. Keeping track of ownership on pieces of equipment, and condition of equipment saves time and money.

Following these steps, ensure well-informed purchases are made that will assist business success in the long term.

# Leasing Hardware as a Service (HaaS)

Leasing computer equipment, also known as Hardware as a Service (HaaS), is becoming a popular alternative to buying new equipment. HaaS offers businesses several advantages, from cost savings to simplified maintenance.

The most obvious advantage of leasing computer equipment is cost savings. Leasing requires an initial payment and then regular monthly payments for the duration of the lease, often at a lower rate than buying the same item outright. These low payments free up capital for other needs, such as employee salaries or additional marketing investments. Furthermore, leases are usually accompanied by customer service packages covering repairs and replacements when necessary. Leasing may reduce associated costs with managing and maintaining software and hardware.

Another benefit of leasing computer equipment is scalability. Businesses can easily adjust technology needs by adding or removing leased items as needed without purchasing them outright. Leasing makes it easier for companies to expand or reduce their operations quickly and efficiently without having to invest in expensive new systems each time changes are needed.

Finally, leasing computer equipment enables businesses to stay up to date with the latest technologies without paying full price all at once. Signing long-term leases allows companies to keep technology current while maintaining cash flow and cost management advantages. This can help ensure that businesses utilize the newest features and capabilities to remain competitive in their respective industries.

Overall, there are benefits associated with leasing computer equipment rather than purchasing. In addition to cost savings and scalability advantages, businesses can enjoy efficiency due to having access to the most recent technologies available on the market. This benefit is without needing to commit significant resources upfront for hardware purchases or paying for costly upgrades whenever new versions become available.

## Software as a Service (SaaS)

Due to its numerous benefits, Software as a Service (SaaS) is becoming an increasingly popular alternative to purchasing software. SaaS offers many advantages over traditional software purchases. Some of the advantages are cost savings, streamlined maintenance, and easier scalability.

One of the main benefits of SaaS is cost savings. Rather than purchase expensive licenses for each user or device that requires access to the software, businesses pay a single subscription fee granting access to all users and devices. SaaS enables businesses to manage IT expenses effectively and in turn save money. Additionally, because companies do not need to purchase hardware or additional licenses to support the software, they reduce upfront costs associated with these components and put those funds towards other investments.

Another advantage of SaaS is simplified maintenance. Instead of having to orchestrate regular updates and patches for on-premises installations, these activities are done automatically by the service provider in a cloud environment. This automation makes it easier for businesses to keep applications up to date without needing dedicated resources for this task. Furthermore, any compatibility issues or bugs identified in the software are expeditiously addressed with no need for manual intervention from IT staff.

Leveraging SaaS enables businesses to scale operations quickly and efficiently without investing heavily in new software licensing agreements each time a change occurs. Companies simply add or remove users from their subscription plan as needed while maintaining access rights and features across different devices and users. This provides businesses maintain full control over operational costs while responding effectively when changes occur in their respective markets or industries.

In summary, there are advantages associated with utilizing Software as a Service instead of purchasing software outright. Companies save money on upfront costs associated with licensing fees, hardware purchases, and additional maintenance activities. Additional benefits include streamlining management tasks for keeping applications up to date, and easily adjusting system capacity according to changing needs without making large investments each time adjustments are required. Overall, SaaS allows companies greater flexibility in how they utilize technology solutions at a reduced cost when compared with traditional methods of acquiring applications through on-premises installations or license purchases.

## Financing for Purchasing, and Project Labor

Many Managed Service Providers (MSPs) offer a variety of financing options to customers to help them acquire the necessary hardware, software, and labor needed to complete their projects. Businesses need to understand the distinct types of financing available so that they can select the most suitable one for their organizational needs.

One common option is purchasing through financing. With a finance option agreement, businesses purchase

IT equipment from MSPs with an agreed-upon payment schedule. This can be beneficial for companies that need specific hardware or software but do not have the funds to purchase it outright. Additionally, because most purchases are short-term agreements, (36-60 months) businesses can upgrade whenever new versions or models become available by staying within a similar payment. This type of financing typically involves lower initial costs over purchasing new technology outright and usually offers flexible payment terms.

There is also the option to purchase hardware and/or software directly from MSPs. Businesses can often take advantage of discounts and other incentives when buying directly from a service provider rather than going through a third-party vendor or reseller. Additionally, many MSPs offer extended warranties on purchased systems which can provide extra protection in case something goes wrong during the life of the product.

Finally, many MSPs offer finance options related to the associated labor costs with specific projects or tasks. Completing this type of finance arrangement typically involves companies agreeing to pay a set fee upfront and then making installment payments as project milestones are completed. This allows businesses to spread payments over time while still being able to receive services as needed instead of waiting until all necessary funds have been acquired.

In summary, there are different financing options available through MSPs providing businesses with flexibility when acquiring hardware, software, and labor needed for projects or other operational goals. Understanding these different forms of financing enables companies to better ensure adherence to budget constraints while ensuring that quality services and technologies are received when required.

# Conflict & Resolution

## Manage Conflict & Resolution

Conflicts are an inevitable part of life in relationships and business. Conflicts arise when there are differences of opinions, needs, or interests. This often creates a situation of disagreement, tension, or even hostility that can be difficult to manage and resolve.

Conflict resolution requires involved parties to fully understand their own perspective as well as the perspective of the other person and work together to produce a solution that is agreeable to both parties. MSPs are in a unique position to guide clients through the process of conflict resolution, offering strategies and advice on how to address an issue and find common ground.

The first step in managing conflict resolution is for your MSP to understand the cause of the conflict and contributing underlying issues. Once established, your MSP can collaborate with both parties to devise a plan for addressing the issue and finding a resolution. This typically involves working to develop an agreement or compromise that meets the needs of all, while considering any potential risks or liabilities. Finally, your MSP can help parties come to a final resolution and ensure agreement with the terms of the solution. Helping clients navigate through conflicts, MSPs play a significant role in maintaining healthy relationships with customers while avoiding costly litigation or further disputes. MSPs should make sure they are familiar with conflict resolution techniques and be prepared to apply them in their interactions with clients.

In addition to offering conflict resolution guidance, an MSP can help parties in understanding the vital role of communication and collaboration as tools for resolving disputes. Encouraging open dialogue between involved parties helps identify the root cause of a particular conflict while developing agreeable solutions. Additionally, working together to create a respectful and productive atmosphere helps resolve differences. Finally, if needed, your MSP can serve as a mediator or arbitrator.

Guiding clients through the conflict resolution process, MSPs help ensure healthy and productive client relationships. With a clear understanding of how to manage conflicts along with strategies for finding successful resolutions, an MSP can serve as an invaluable partner in conflict prevention and resolution.

## Conflict Between MSP and Client

If a conflict arises between the MSP and their client, both parties need to communicate openly about their needs and expectations. The MSP should also be willing to listen to the client’s concerns and work with them to find a working solution. It may be beneficial for the MSP to offer alternative solutions or services that effectively meet the client's needs. If all else fails, an unbiased mediator or third-party arbitrator may be able to help parties agree without further escalations of tensions or damage to the relationship. Ultimately, taking steps to manage conflicts as soon as they arise, MSPs ensure healthy relationships with their clients and provide the best service possible.

# Changing MSPs

When an organization needs to change a Managed Service Provider (MSP), it is important to take the time to evaluate the options before deciding. Knowing in advance what services your current and potential MSPs offer can be key in finding the best fit for an organization. Additionally, organizations should assess the capabilities of each MSP, including technical and customer support expertise, as well as scalability, security protocols, and service level agreements. It is important to review pricing structures and related fees such as setup fees or cancellation fees when comparing MSPs. Furthermore, organizations should research feedback provided by past customers. Customer feedback helps identify issues encountered with a given service provider. Additionally, understanding contractual terms associated with switching providers allows organizations to anticipate any costs or other obstacles that could arise during the transition process.

Communication between the existing MSP and the incoming provider is critical for ensuring a smooth transition process. Both parties should understand what needs to happen for data and systems to be effectively transferred from one provider to another. Organizations should consider if there are legal or privacy requirements needing to be addressed upon switching providers. Timelines for completing the transfer of services and systems is another consideration. Taking the time to understand these requirements ahead of time ensures a successful transition and cost-effectiveness.

## Backup the Data

Having a valid backup of data on both the cloud and an external device such as a USB drive before beginning migration processes is a key step to ensuring that your organization's information remains secure. Not only does backing up data help protect against potential disasters, but it ensures that if something does go wrong during the transition process, you will have access to all important documents and files. Backups provide an additional layer of security against malicious actors attempting to steal or corrupt data. Backing up your data regularly streamlines the migration process making it easier to restore files quickly and accurately in the event of disruption associated with transitioning from one MSP to another. Having a valid backup strategy in place provides peace of mind for involved stakeholders with the process due to secure and protected information in the event of an unsuccessful migration.

When creating a backup strategy, organizations should consider how often to back up data. Ideally, back-ups should be daily or more frequent depending on the size and complexity of the project. Additionally, organizations should make sure that backups are securely stored offsite to protect from potential damage or corruption caused by natural disasters such as flooding or fires. Organizations should consider any General Data Protection Regulation (GDPR) compliance requirements when choosing a storage location for their backups. Utilizing encryption technologies help safeguard sensitive company information from unauthorized access through the encryption of sensitive data that cannot be read without permission from authorized personnel. Finally, having a comprehensive testing plan in place ensures backups are working properly across all systems when restoring them later. Again, this saves time and money for businesses when migrating.

## Give a Chance to Your Current MSP

It is important to give your current MSP a chance to modulate the transition. Research any feedback from past customers and understand any contractual terms that arise with switching providers. Talk to both the existing and incoming MSPs for a comprehensive understanding of what needs to be done. Be sure to follow any legal or privacy requirements when moving data. Back up data before beginning the transition process so it is safe even if something should go wrong. Provide your current MSP an opportunity to ensure everything can go smoothly and quickly.

# Contract and Agreement Guide

## Contracts

Contracts are important because they provide legal protection and serve as an enforceable shield of safety for both parties involved in an agreement. Outlining all terms, conditions, and expectations in a contractual agreement allows both parties enter the relationship confidently without feelings of uncertainty or lack of protection. Contracts allow both sides to have a clear understanding of the rights, responsibilities, duties, and liabilities, essential to conduct any kind of business transaction. Having a contract avoids disputes that may arise since it provides proof that all points have been explicitly mentioned and agreed upon by both parties.

## Question of Contract Necessity

Having a contract with an MSP is beneficial in many ways. Most importantly, a contract provides clarity and security to both parties. By signing a contract, both the MSP and the client know exactly what services are being provided, as well as the rules governing those services. This helps prevent any misunderstandings about what is expected of either party. Additionally, having a contract allows for recourse if either party fails to adhere to its agreed upon obligations.

Moreover, having a written and agreed upon service-level agreement (SLA) ensures that each party is aware of their rights and responsibilities related to performance and customer satisfaction guarantees. An SLA sets expectations for response times and resolution times for issues that may arise. An SLA ensures customers receive the level of service they expect from an MSP, while keeping costs under control.

Contracts protect legal issues arising out of an agreement between the two parties. A contract outlines exclusions or limitations to services provided by the MSP so there is no confusion about what is covered under the agreement; this specificity protects both parties in case of litigation or arbitration proceedings. Likewise, contracts provide accountability on both sides since they define payment terms which include when payments are due from the client and acceptable methods of payment such as credit cards or electronic funds transfers (EFT).

Therefore, having a contract with an MSP can provide protection against legal issues, set clear expectations for customer satisfaction and performance levels, as well as ensure appropriate payment terms are established upfront—all reasons why you should have one in place.

## Contracts & Agreements

Contracts and agreements, while similar in certain respects, operate under two distinct and separate frameworks. Contracts are legally binding documents that determine the particulars of any given relationship or transaction between two or more parties; a contract is authorized by law and if breached may be enforced in a court of law. Agreements on the other hand do not hold the same weight - they serve as informal outlines of expectations that both sides must adhere to but cannot necessarily be held accountable for should one side fail to meet them.

Unlike agreements, contracts are often more detailed

and require signatures from all participating parties. Additionally, they state what will happen in the event of a violation by either side. For instance, if someone fails to obey the terms of the contract then they may be held accountable for damages depending on specific occurrences. On the other hand, agreements tend to have fewer details concerning expectations and potential ramifications for non-fulfillment; these documents also typically take less time to create.

A further distinction between contracts and agreements is the purpose they serve. Contracts are typically drafted to establish a lawful bond with laid-out expectations from both parties, while agreements aim to outline shared objectives without imposing any legal requirement on either side.

In the end, contracts frequently involve sophisticated financial terms such as payment deadlines or customer satisfaction guarantees. Agreements do not consist of complex monetary arrangements but rather contain more broad language regarding obligations made between the two parties instead of liabilities that may emerge in case either party fails to meet their commitments.

# Resources to Help Choosing an MSP

## Best Practices on Selecting an MSP

On the hunt for an ideal Managed Service Provider (MSP)? Making a choice that suits you can be bewildering. If your IT knowledge is limited, engaging a consultant to assist with selection may prove worthwhile. Moreover, it is prudent to recruit an audit expert who has no personal stake in your MSP; this method ensures obtaining yearly evaluations conducted by neutral parties. These assessments are devised to offer impartial feedback on how effectively your MSP operates and functions.

## Benefits of Hiring an MSP Auditor

Hiring an MSP auditor yields several benefits to organizations, which is why it is often recommended.

The first benefit is that the auditor provides a more objective and independent assessment of the provider. An auditor will take a closer look at how the MSP operates and functions, as well as their processes, policies, and procedures. This review enables organizations to make informed decisions before committing to any partnership with an MSP. In addition, regularly scheduled audits guarantee that reviews are kept up to date with current standards and best practices.

The second benefit is that auditors can spot potential security risks in advance. Auditors are experts in detecting vulnerabilities in the system and identifying areas for improvement. With their help, organizations can proactively address issues. As a result, companies ensure that they are taking steps to protect their data and privacy while also complying with industry regulations such as GDPR or HIPAA.

A final benefit of MSP auditors is the feedback provided regarding the technical capabilities of the MSP. Auditors will look at various aspects such as pricing models, customer service levels, backup and disaster recovery procedures, and scalability options, so organizations have a better understanding of what they should expect from the MSP. This enables companies to select one that meets their requirements while simultaneously providing value for money.

Overall, recruiting an MSP auditor is essential for every organization looking for a reliable managed service provider who offers quality services at competitive costs. The auditor provides valuable insight into how effectively the provider operates along with feedback about its technical capabilities allowing organizations to make more educated decisions.

## Where to Look to Find a Good MSP

Finding a good Managed Service Provider (MSP) can be a daunting task if you do not know where to look. Before deciding on an MSP, it is important to do some research to ensure valuable experience, qualifications, and certifications. Here are some of the best places to start your search:

**Online Reviews & Ratings**
Many managed service providers have online reviews and ratings that help make an informed decision. Take the time to read customer feedback and find out what previous customers thought about their experience with the provider before deciding. Pay attention to both

positive and negative reviews so you can accurately gauge the quality of their services.

**Professional Associations & Organizations**
Professional associations such as ITIL or ISACA provide organizations with access to valuable resources that can help choose a reputable MSP. Joining these associations provides organizations with access to industry standards and best practices along with recommendations from other members who have worked with a particular MSP in the past.

**Industry Publications**
Industry publications such as trade magazines and journals are another major source of information when choosing an MSP. These publications often feature articles written by MSP experts who provide useful insights into assorted topics related to managed services, including security protocols, cloud computing, data analytics, and disaster recovery solutions. This type of content enables organizations to gain a better understanding of what they should expect from an MSP before making any commitments or investments in their services.

**Networking Events**
Participating in networking events is another effective way for companies to find well-managed service providers in their area or field of interest. Attending conferences or tradeshows provides companies with the opportunity to meet different MSPs in person and ask questions about their solutions and capabilities allowing them to better vet potential providers.

There are various sources businesses can turn to when looking for a well-managed service provider (MSP). Doing research beforehand will ensure that you select one that meets your specific requirements while providing value for the money. In addition, hiring an external auditor may be beneficial as they offer independent assessments giving organizations peace of mind before finalizing partnerships with providers.

## Hiring an MSP for a Trial Period

Yes, it is possible to hire a managed services provider (MSP) on a trial basis. Depending on the provider, this may involve an initial setup fee as well as an agreement for the duration of the trial period.

Typically, companies use this period to assess their needs and determine if the MSP's services are a suitable fit for their business. During the trial period, most MSPs will guide how to best use their services, allowing companies to assess distinctive features and configurations before committing to any long-term contracts.

Most providers offer discounts or incentives during the trial period that can be used to offset some of the costs associated with setting up and evaluating their offerings. This provides an opportunity for companies to explore options to find the solution that meets their needs.

An essential component to hiring an MSP on a trial basis is having access to reliable support throughout this process. As such, make sure to choose one that offers 24/7 customer service and technical assistance should help be needed with any aspect of the trial or should there be a need to discuss changes or upgrades. In addition, check what type of reporting and analytics are offered so you can monitor your usage and performance during this trial period.

Overall, there are benefits associated with hiring an MSP on a trial basis before committing to any long-term partnerships. If available, companies should take advantage of this option to evaluate different solutions and providers without committing too much money upfront. With thoughtful planning and due diligence during the selection process, businesses can ensure they select a managed services provider that meets their requirements while providing monetary value.

# Thoughts on Selecting an MSP

When selecting a managed services provider (MSP), it is important to ensure the provider meets your organization's requirements and provides value for money. It is essential to have a proper partnership with an MSP to ensure the success of your business operations. An MSP should be able to provide the expertise and resources needed to meet the demands of your IT infrastructure.

Having an established partnership with an MSP helps businesses stay ahead of competition by providing access to innovative technology and solutions. An MSP typically has experience in various areas such as cybersecurity, cloud computing and data analytics to name a few. This allows organizations to leverage the expertise of an MSP and take advantage of emerging trends and technologies. Additionally, having a dependable partner helps manage business operations reducing operating costs and maximizing efficiency without compromising security or data integrity.

In addition, building a long-term relationship with a reputable MSP creates trust, making it easier to communicate expectations. A good partnership should include regular communication protocols and feedback loops so that issues or challenges are addressed promptly. Communication ensures the MSP stays up to date on innovative technologies and trends to provide tailored recommendations related to organizational goals or objectives.

When it comes to managing their IT infrastructure selecting an experienced managed services provider is critical for businesses looking for expert guidance. Companies should conduct thorough research before making any commitments or investments to find a suitable partner who understands their individual needs while providing value for money. Doing so will not only help improve operational efficiency but also create a partnership built on trust, transparency, and mutual respect which will benefit all parties.

# Biographies

## Dr. Julia Diakakis

Dr. Julia Diakakis, a highly respected and decorated educator and businesswoman, boasts an impressive career spanning over 35 years. Born into a military family in Bremerhaven, Germany, Dr. Diakakis was raised throughout Europe providing her a diverse upbringing which laid the foundation for a versatile professional journey.

A proud graduate of Ohio University, she holds a B.S. in Education with a minor in English. Her certifications span various subjects including English, Math, History, Social Science, Earth Space Science, and Elementary Education with endorsements in Reading and English Speakers of Other Languages (ESOL). Her passion for education led her to different roles in educational institutions ranging from elementary to high school levels, earning her recognition as one of Florida's most esteemed educators. Furthering her passion for education, she has served on dissertation committees as the content expert for doctoral students.

Parallel to her educational career, Dr. Diakakis has been maintaining an IT business since 1994 and served the Brevard School Board as an IT Specialist for seven years. This combination of education and technology expertise uniquely positions her at the intersection of these two critical fields.

Dr. Diakakis further expanded her skill set by obtaining a master’s degree in educational administration, leading to her principal certification and an administrative role for about 15 years. She didn't stop there; her boundless passion for education, business, and information technology propelled her to pursue a Doctorate in Organizational Leadership and Development.

A member of prestigious societies such as Alpha Chi Honor Society, Kappa Delta Pi Honor Society, and National Society of School Administrators, Dr. Diakakis has made significant contributions to developing computer lab programs for Brevard Public Schools in Florida. She also pioneered the first fully operational Discourse Study Computer classroom in Florida. Her extensive work with numerous private companies throughout her career testifies to her business acumen.

Dr. Diakakis' unique blend of educational knowledge and business insights makes her a formidable force in her field. Her life and career serve as an inspiration for many who aspire to make a difference in the realms of education and business.

## Contact

**E-mail: authors@diakakis.us**

**Website: https://diakakis.us**

# Saki Diakakis

Born in Athens, Greece, Saki Diakakis embarked on his journey with a classic high school education followed by military service as an Amphibian Green Beret. In 1989, Saki immigrated to the United States to nurture his passion for aviation and began training as a private pilot, initially aiming for a career as a commercial pilot.

However, he found himself drawn towards the burgeoning field of computers, networks, and software development. Saki returned to academia, this time immersing himself in computer science, and subsequently ventured into diverse IT industries.

From the early 1990s, his career path led him to work with local city government bodies as a Network Systems Analyst and Software Developer. His professional portfolio expanded across multiple sectors including county, city, and state governments, in areas of advertising, space subcontractors, home builders, and water treatment plants, among others. During his time with city government he designed and built the TV station for the City of Cocoa Beach, marking a significant milestone in his career.

During a nine-year stint in the advertising industry, Saki managed central networks and developed integral systems for the national yellow page company. As a consultant, he contributed to data analysis and application development. Over the years, he has developed an array of commercial applications encompassing document management systems, CRM, billing, building department permits, fleet management, and more. Adding another feather to the cap, he pioneered one of the first surveillance wireless undercover applications in the United States.

For the past 12 years, Saki has been associated with a local Managed Services Provider, specializing in computer networks, telephony, and cybersecurity applications. His services have benefited hundreds of businesses on local, federal, and international levels.

Saki's extensive formal education and practical experience have equipped him with knowledge that many businesses and stakeholders consider invaluable for their success. He is committed to leveraging this expertise to continue the contribution to the field of technology and business.

**Contact**

**E-mail: authors@diakakis.us**

**Website: https://diakakis.us**

# NOTES

# NOTES

Made in United States
Orlando, FL
04 November 2026

90207803R00069